99 Ways to Improve Your Shortwave Listening

by

Len Buckwalter

HOWARD W. SAMS & CO., INC.
THE BOBBS-MERRILL CO., INC.
INDIANAPOLIS · KANSAS CITY · NEW YORK

FIRST EDITION

FIRST PRINTING—1972

International Standard Book Number: 0-672-20911-X
Library of Congress Catalog Card Number: 72-76099

Preface

Shortwave pioneers at the turn of the century had remarkable receiving instruments. With little more than a coil of wire, a simple detector, and earphones, Guglielmo Marconi in Canada heard the first transatlantic signals from England in 1901. The fact that he raised his antenna to a height of several hundred feet on a kite during a gale doesn't make the feat any less astounding. Radio waves have the ability to rise some 100 miles high, reflect off the ionosphere, and return to the earth after spanning continents—and suffer little loss of strength along the way. After more than a half-century of progress, which has seen the introduction of tubes, transistors, integrated circuits, and other developments, you would think there is little room left for improving the art of shortwave listening.

That is hardly the case. Whether you own a pocket portable or a deluxe communications receiver, there is endless opportunity for improvement. A large number of variables confront a shortwave listener that can mean the difference between hearing a rare station or trying to tune amid static crashes and other noises, both natural and man-made. An antenna installation alone offers many opportunities for experimentation. There are numerous accessories that make receivers more versatile or more sensitive to weak signals. Listening while you're away from home, knowing how to prepare reception reports that "pull-in" QSL cards, or obtaining special program schedules or propagation reports can all add to your listening pleasure. These, and dozens of other steps you can take, often with little technical skill, can improve the already fascinating hobby of shortwave listening.

LEN BUCKWALTER

Contents

Section 1. Antennas and Installation

Section 2. Receivers

Section 3. Accessories

Section 4. QSLs, Skeds, and Logs

Section 5. Techniques

Section 6. Listening Away From Home

Section 7. Special Signals and Bands

Section 8. Interference

Section 9. Troubleshooting

Section 10. Miscellaneous

SECTION 1

Antennas and Installation

1

A Tuned-Doublet Antenna

Although simple long-wire antennas provide satisfactory shortwave reception, considerable improvement can be obtained by using specialized antennas. A good example is the half-wave tuned doublet (also known as the "dipole" antenna). A tuned doublet performs better than a long-wire antenna for two reasons: (1) a doublet is tuned to only one band of frequencies and captures more of the passing wave at those frequencies, and (2) it displays directional characteristics, which tend to concentrate the passing signal in the manner of a lens. These directional characteristics, however, are also somewhat of a disadvantage in that the doublet may perform poorly at frequencies other than those for which it was designed. Thus, you may wish to construct a doublet for high-efficiency reception on a favorite band, then switch to a long wire for general listening on other bands. (A suitable changeover system is described in Item 2.)

Some typical half-wave doublet dimensions for several of the popular shortwave bands are given in Table 1-1. If you want to build a doublet for a frequency not listed here, use the following simple formula to obtain the correct dimensions:

$$L = \frac{468}{f}$$

where,

L is the physical length of the antenna in feet,
f is the frequency in megahertz.

Table 1-1. Typical Half-Wave Doublet Antenna Lengths

Band (Meters)	Frequency (MHz)	Doublet Length (Feet)
41	7.100- 7.300	67
31	9.500- 9.775	50
25	11.700-11.975	40
19	15.100-15.450	32
16	17.700-17.900	28
13	21.450-21.750	23
11	25.600-26.100	19

The answer (L) is one-half wavelength in feet for the frequency in question, and is the correct length for a half-wave doublet, or dipole antenna.

Details for constructing a half-wave doublet antenna are shown in Fig. 1-1. Note that the antenna length is the total distance between the two end insulators. First, cut the antenna wire to this length, then cut the wire exactly in half for connecting to the center insulator. The lead-in wire may be either tv twinlead or coaxial cable, although for best electrical match, coaxial cable having an impedance of about 75 ohms (such as Type RG-59/U) should be used.

For best performance, try to support the antenna as high as possible. An antenna of this type works best when mounted one-half wavelength (the same as its overall length) above the ground. Lower heights tend to favor the shorter-range signals arriving from higher angles over the horizon. The more distant skip signals from overseas tend to arrive at very low angles, and the antenna should be mounted to favor them.

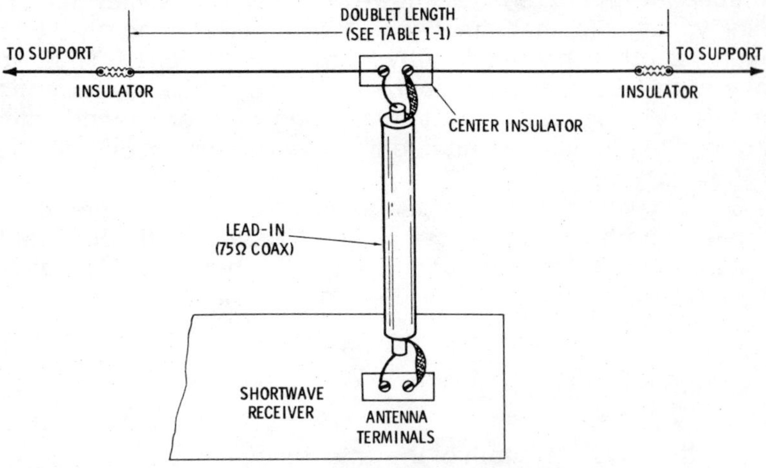

Fig. 1-1. A half-wave doublet antenna.

A half-wave doublet antenna is most sensitive to signals arriving *broadside* to the wire. The ends of the antenna wire point to "nulls," or areas of poor reception. In other words, if you want your antenna to be most sensitive to signals arriving from the east or west, orient the antenna wire in a north-south direction.

2

Doublet/Long-Wire Antenna Selector

If you use a directional (or tuned) antenna such as the doublet described in Item 1, a simple switching system will increase its usefulness. As shown in Fig. 2-1A, a doublet is normally connected to two separate antenna terminals on the shortwave receiver. The same antenna, however, can be converted to a long wire by connecting both lead-in wires to one receiver terminal. This is illustrated in Fig. 2-1B.

To avoid disconnecting and reconnecting the wires at the receiver antenna terminals, you can wire an inexpensive switch to do the same thing. The switch is a common bakelite type widely used for tv antenna switching: a double-pole, double-throw knife switch. Wire it as shown in Fig. 2-1C and you will obtain the changeover system just described.

Be sure to check for the best connection at the shortwave receiver terminals. Some sets have three terminals (one is a

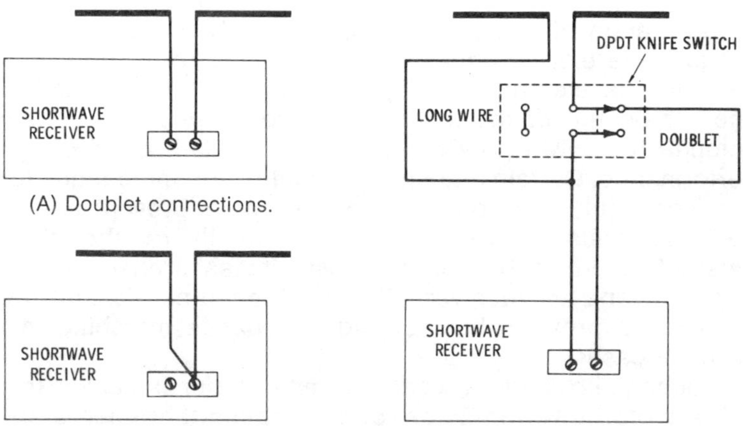

(A) Doublet connections.

(B) Long-wire connections.

(C) Selector-switch connections.

Fig. 1-2. Doublet/long-wire changeover system.

ground connection), while others indicate "A" (antenna) and "G" (ground) on two terminals. Experiment with the connections to these terminals to find the ones which produce maximum signal strength. Then, you can switch to the doublet when listening to the frequency band for which it was cut, and switch to the long-wire antenna when listening to the other frequency bands.

3

Make an Antenna Tuner

An antenna tuner placed between the shortwave antenna and the receiver can improve reception in many ways. First of all, it is a tuned circuit that is quickly adjustable to any major shortwave or amateur band in the frequency range of from 3 to 30 MHz. Thus, it covers the following shortwave bands: 13, 16, 19, 25, 31, 41, 49, and 60 meters. It includes the amateur bands on 10, 15, 20, 40, and 80 meters. Since tuning is continuous, other services between these bands are also included.

Receiver *sensitivity* is especially improved with an antenna tuner in sets that have no rf amplifier. The circuit introduces a tuning element missing in much commercial equipment. *Selectivity* is similarly improved. One advantage of this is that *image* signals entering a receiver that had poor selectivity now encounter a more difficult path. Cross modulation and overload from strong signals away from the desired signal are also minimized.

One of the leading advantages of the tuner is that it improves the operation of short antennas, especially those that are not cut to some exact frequency. The length differences are compensated for electronically by adjusting the tuner controls. Yet another benefit of the tuner is that you may use a variety of antenna types with any receiver and maintain good electrical performance. Certain antennas present a high impedance to the receiver input, while others, like the dipole, present a low impedance. Thus, different antennas normally require different types of lead-in cable for the most efficient transfer of signal from antenna to receiver. This antenna tuner is sufficiently versatile to provide the required impedance matching in any of these cases.

A schematic diagram of the antenna tuner is shown in Fig. 1-3. A pictorial wiring guide for assembling the tuner is shown in Fig. 1-4. Photographs of the completed tuner are shown in Figs. 1-5A, B, and C.

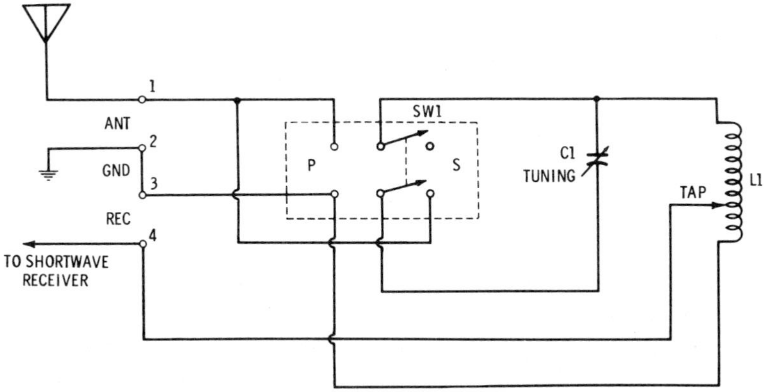

Fig. 1-3. Schematic diagram of an antenna tuner.

Table 1-2. Parts List for Antenna Tuner

Item	Description
C1	Capacitor, variable, 330 pF.
L1	Coil, 3″ long, 1″ dia., 32 turns per inch (B & W Type 3016 or equiv.).
SW1	Switch, slide, double-pole, double-throw.
	Perforated board, 6″ x 3″.
	Fahnestock clips (4).

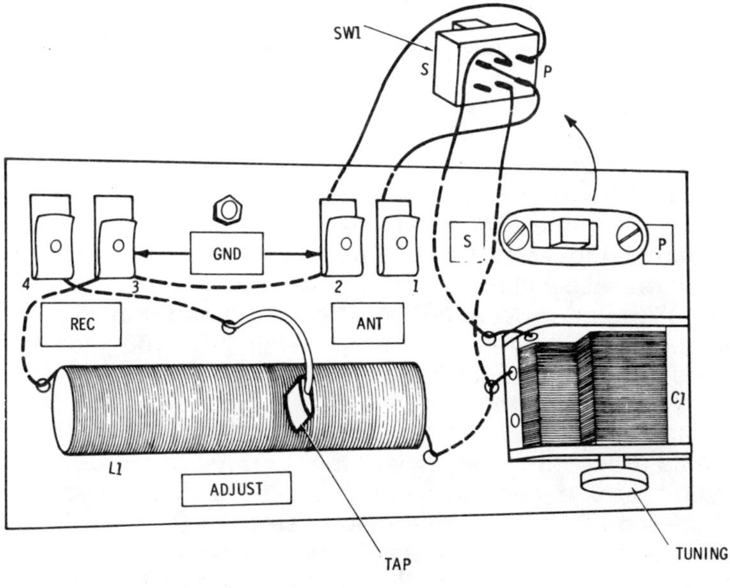

Fig. 1-4. Pictorial wiring guide for Fig. 1-3.

As seen in the photographs, the tuner is assembled on a piece of masonite pegboard, although perforated phenolic board will do just as well. You can wind the coil yourself if you wish, but it is usually more convenient to purchase a commercial unit (see the parts list in Table 1-2). Mount some type of supporting feet at the corners of the board (machine screws will do, but recessed rubber bumpers are best for this) to prevent the underside wiring from short-circuiting against a metal surface. The item shown as the "tap" is a solder lug or other piece of thin metal that can be inserted between the coil turns for adjusting the coil (the tap is adjusted when you change bands).

When the tuner is completed, place it on or near the receiver (see Fig. 1-5C). Connect the receiver antenna terminals to terminals 3 and 4 (REC) on the tuner, using a pair of short, twisted wires. Note that terminal 3 is a ground terminal, so be certain it is connected to the ground terminal on the receiver. (If your receiver has no ground terminal, just connect terminal 4 on the tuner to the antenna terminal on the receiver.) Next, connect the lead from the external shortwave antenna to terminal 1 (ANT) on the tuner. If the antenna has two leads (as with shielded lead-in), connect the shield to terminal 2, the other ground terminal on the tuner. Connect one of the tuner ground terminals (2 or 3) to a good external ground such as a cold-water pipe.

Now, while operating the shortwave receiver on one of the major bands, make the following adjustments on the tuner: First, set the tuning capacitor (C1) to its midposition (plates halfway meshed). Remove the tap from the coil and slide it along the turns from one end to the other. At some point along the turns there should be some indication that you have found the proper tuning range for the frequency band your receiver is switched to—either an increase in background or noise level, an increase in volume of a station signal, or, if your receiver has an "S" meter, a peak indication on the meter. Since your hand on the tap may detune the circuit slightly, you may have to make this adjustment through trial and error. When you find the point on the coil that produces a peak indication for a particular band, mark the location point for future reference.

This procedure enables you to determine the proper number of coil turns needed to resonate each band. The best position of the tap is the point where you can peak the signal (produce a rise and fall in signal strength) by rocking the tuning capacitor back and forth. Consider the tap as the coarse tuning, and the tuning capacitor as the fine tuning for each band.

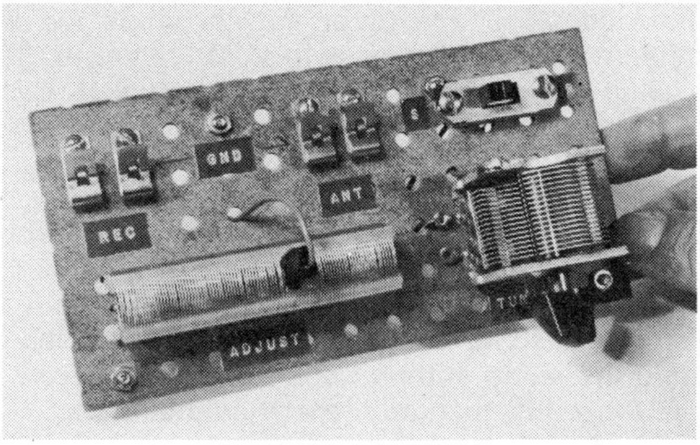

(A) Top view.

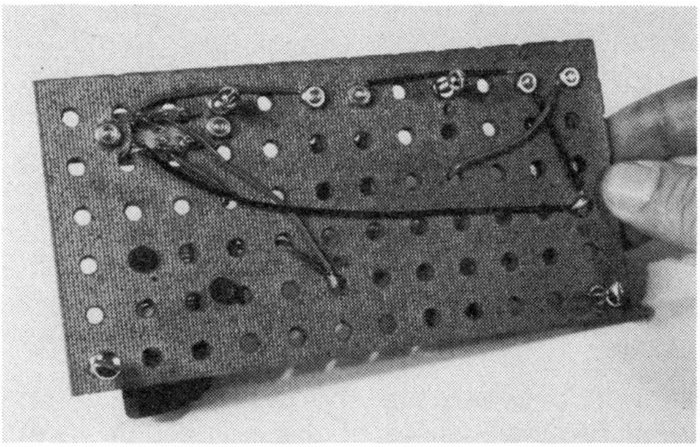

(B) Bottom view.

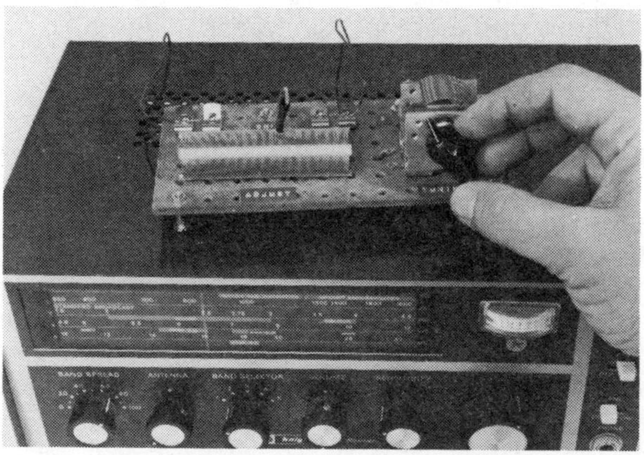

(C) Tuner in use with shortwave receiver.

Fig. 1-5. Photographs of the completed antenna tuner.

On some occasions, the tuner may produce the wrong results; i.e., it may cause a decrease in signal strength. Should this occur, it simply means that the series-parallel slide switch (marked "S" and "P" on the illustrations) is set to the wrong position. This switch (SW1) is in the circuit so that the coil (L1) and the tuning capacitor (C1) can be connected either as a series-resonant circuit or as a parallel-resonant circuit. Since the tuner may have to work with a variety of antennas, you can instantly select the position that produces the best electrical match. You may also have to change the switch when changing bands, since the same antenna operating on different frequencies may require this change in electrical matching.

Finally, when the tuner is operating satisfactorily, experiment with the ground connection. On some frequency bands, reception might be improved by disconnecting the ground lead between the tuner and the receiver, or the ground to the cold-water pipe. Some experience with the circuit will help you determine the most efficient connections for your particular installation.

4

Low-Resistance Indoor Ground

A connection between the shortwave receiver chassis and the electrical ground system of the home has several advantages. The antenna system may operate more efficiently, interference from appliances is more effectively shielded by the receiver cabinet, and there is an increase in lightning protection. To gain these advantages, the connection should be made with the lowest possible electrical resistance between the receiver and ground. There are several ways to do this job.

Start by locating the grounding point in your receiver. In many cases this is labeled, or is indicated by the manufacturer in the instruction manual. If the receiver is an ac-dc type, it is especially important to only use a terminal marked "ground" to prevent short circuits. In solid-state receivers, avoid connecting to the heat sinks that are used to cool the output transistors. These are metal plates with fins that increase the surface area, and are usually not connected to the ground point in the receiver. When you have located the ground point, connect a heavy wire to it. Use at least No. 18 wire, the size used for lamp cord. Run the wire to the nearest ac outlet and connect it to the screw that fastens the outlet cover plate to the outlet box. If

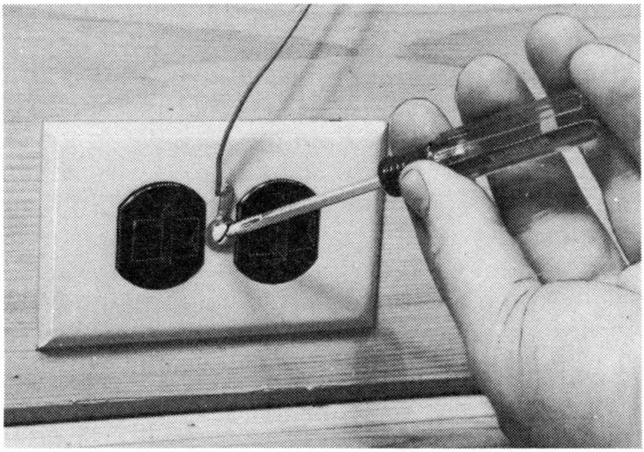

Fig. 1-6. Making a low-resistance indoor ground connection.

your electrical wiring was done according to the National Electrical Code, this screw connects to the electrical ground system of your home. Fastening the ground wire to the screw is easier done if you first attach a solder lug with a hole in it to the end of the wire. The lug will then be retained by the cover screw (see Fig. 1-6).

Another method of grounding is to connect the grounding wire to a cold-water pipe using a suitable metal clamp. Although this is often recommended, it may have some drawbacks. For example, there may not be a pipe close enough to your installation, and you would then have to run an unnecessarily long ground lead. Also, because nonmetallic piping for water systems, and insulating couplings on metallic piping systems are being used more and more, the water piping in your house may not be electrically continuous to the outside, or "earth" ground.

You should *not* connect your ground lead to a gas pipe because gas piping systems are often constructed with insulating joints, or are of coated or nonmetallic piping. Also, the use of gas piping for electrical grounding is not permitted by local authorities or gas suppliers in many parts of the country.

5

Earth Ground

An excellent method of grounding your receiver and antenna system is to run the grounding wire directly into the earth. If this is practical in your location; i.e., if the grounding wire does not have to be unnecessarily long (as it would have to be, for example, if you live several stories above ground in an apartment building), it is best to consider this method. To be most effective, the grounding wire from the receiver to the earth electrode should be at least No. 10 solid copper wire with no insulation, or No. 8 solid aluminum wire. The grounding wire should be run by the most direct route to the nearest underground water piping system on the street side of your water meter, or to a cold-water pipe as near as practicable to the water service entrance to the building. If your source of water is from a driven well, the connection should be made as near as practicable to the well. (The pipe must be metallic, of course; nonmetallic piping is unsuitable.) Connection to the pipe is made with a special grounding clamp available where electrical supplies are sold.

If there is no suitable water pipe near your installation, you can use what electricans term a "made electrode." One type of made electrode commonly available is a ½-inch copper rod that is driven into the earth, preferably below the moisture level. The recommended length for a ground rod is 8 feet, but this may not be practical if you live on rocky soil. When rock obstructs the rod, the rod may be buried as deeply as possible in a horizontal trench. The grounding wire is fastened to the top end of the rod with a screw terminal.

6

Low-Noise Antenna

Here is a simple antenna you can assemble that will help reduce the pickup of "noise," or electrical interference from man-made sources. These interference sources include electric motors, automobile ignition systems, power lines, and home appliances. This type of interference tends to remain close to the ground and not radiate upward. Many shortwave antennas are merely long wires, the vertical, or lead-in portion of which must often pass through strong interference fields, and thus will pick up excessive noise from these fields. This can be prevented by using coaxial cable for the lead-in portion of the antenna. Coaxial cable has an inner conductor surrounded by a braided-wire outer conductor that acts as a shield for the inner conductor. The shield is grounded, and thus virtually eliminates noise pickup by the inner conductor. The sensitive, or horizontal, portion of the antenna is mounted high enough above ground so that it may pick up maximum signal and minimum noise.

Another advantage of coaxial cable is that it does not have to be kept away from metal along its run. It can, in fact, be run underground. One precaution should be observed, however, and that is to keep the coaxial cable as short as possible. An excessive length of lead-in may cause signals of certain wavelengths to be trapped or weakened within the cable. This is not a problem when the antenna is cut for a specific frequency and electrically matched to the coaxial cable. In this low-noise antenna, however, one length of antenna wire must operate on many frequency bands, and an excessively long cable becomes a problem.

The type of coaxial cable to use is not too critical. You can choose from several common types that are sold for Citizens-band or television transmission line. The impedance of these cables ranges from about 50 to 75 ohms.

Before connecting the cable to the antenna, strip away about 2 inches of the outer jacket and braided shield from one end. Then solder the inner conductor to one end of the horizontal antenna wire. As seen in Fig. 1-7, a length of 30 to 100 feet is indicated for the antenna, but the longer it is, the better it will perform. Both ends of the antenna wire must be connected to

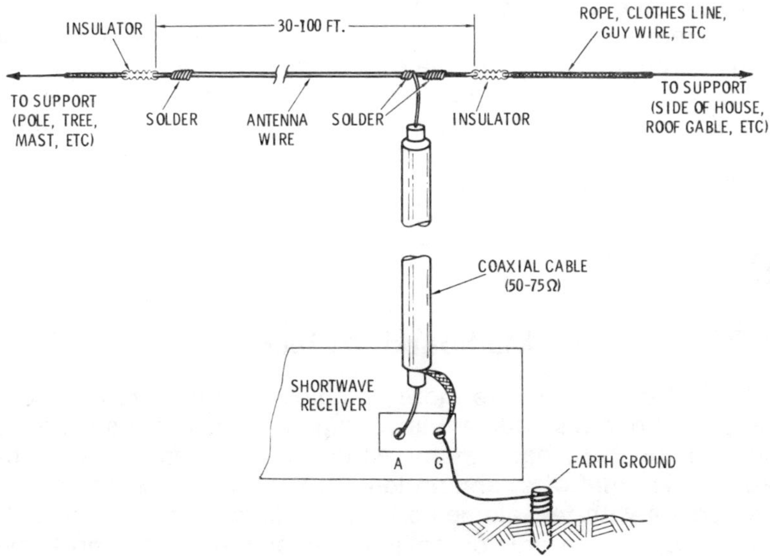

Fig. 1-7. Low-noise antenna.

insulators, and then strung between two physical supports (such as between the side of the house and a tree, for example). This is to keep the antenna insulated from ground and also to relieve the physical strain on the antenna wire where it connects to the lead-in cable.

Connect the other end of the coaxial-cable lead-in to the receiver antenna terminals. As shown in Fig. 1-7, the inner conductor and shield go to terminals "A" and "G" respectively. This system will work to best advantage when the ground terminal is also connected to earth ground in the electrical system of the home as discussed in previous items.

7

Telephone Antenna

This trick does not always work, but it's worth a try. Connect a short length of antenna wire between the shortwave receiver and the fingerstop of a telephone (the curved piece of metal that stops the finger as you dial). In some instances, you will discover excellent reception on some bands. It may not prove to be the most practical antenna for regular listening, but is an interesting novelty.

Why does it work? One guess is that the telephone fingerstop is connected to an electrical ground which might be dozens of feet away. The run of wire between the fingerstop and actual ground, therefore, may form an antenna section which tunes (resonates) certain wavelengths. Since the signal voltage of a tuned antenna drops to zero at certain points (null points), the tuned frequencies are not short-circuited by the ground connection. (Anyway, it works . . . sometimes.)

8

Multiband Antenna

This antenna is a little more complex than the ordinary long wire, but it offers significant advantages. Since it has a separate element tuned to each major band, antenna efficiency is improved over that of a random length. You may construct a simple version with few elements to cover favored bands, or build a more elaborate version to tune all of the major shortwave bands. The limit is the amount of space you have available. The

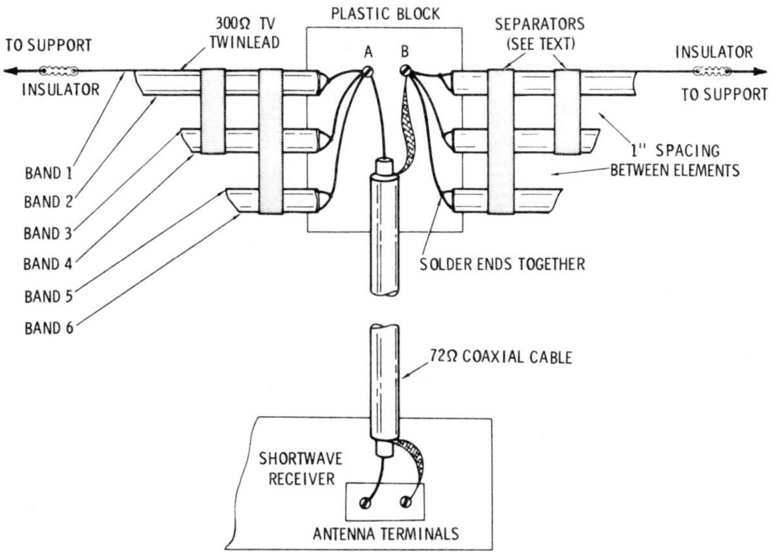

Fig. 1-8. Multiband antenna.

lowest frequency band (longest wavelength) requires an antenna length of 67 feet.

Two features of this design should prove attractive to the builder. First, it requires no loading coils or bandswitches to change bands, as in other multiband antennas. Each band has its own dipole element permanently connected to a coaxial lead-in. The elements are connected to common points, but they are electrically isolated through differences in tuning frequency. Secondly, the antenna is made from standard 300-ohm tv twinlead—the flat-ribbon type. The advantage of this is that each length of twinlead can serve as two antennas since each length contains a pair of wires separated by a plastic insulator. Note in Fig. 1-8 that three horizontal twinlead lengths create a 6-band antenna.

Although the antenna described here has dimensions for the six most popular shortwave bands, you are not limited to these frequencies. After you learn the basic procedure, the elements may be cut to any desired band. The following description, though, should take care of most situations.

Note in Table 1-3 that Bands 1 and 2 are grouped together since they are formed from the same length of twinlead. Always start by placing the lowest frequency band (which corresponds to the longest wavelength in meters) at the top position of the antenna. As the longest element, it will support narrower lengths

Table 1-3. Overall Antenna Lengths for Six Popular Shortwave Bands

Band No.	Wavelength (Meters)	Overall Antenna Length
1	41	67
2	31	50
3	25	40
4	19	32
5	16	28
6	13	23

suspended in the lower positions. In this example, the lowest band is 41 meters, which calls for an overall length of 67 feet of twinlead. This forms the topmost element. If you shorten the lower copper wire inside the same length of twinlead to 50 feet, you form an antenna for Band 2, or 31 meters. After the wires are cut to size, fold each overall length of twinlead in half to locate the center point, and cut both wires. As shown in Fig. 1-8, the two wires from each twinlead half are soldered together and connected to center terminals "A" and "B" on the plastic block. (The block serves as the support and connecting point for all wires at the center of the antenna.)

You can start Bands 2 and 3 with a second length of twinlead cut according to the dimensions given in Table 1-3. The second twinlead element is spaced an inch below the top element. These elements are kept apart and supported by plastic electrical tape, or pieces of solid plastic (shown as "separators" in Fig. 1-8). Estimate how many separators your antenna needs, as this will vary according to element size. The topmost element, which supports the whole array, is tied to end supports through insulators.

Do not connect the outer ends of the twinlead together—they must be electrically free. All the elements are joined at the center point only, where the shield and center conductor of the coaxial cable are connected to points "A" and "B." Use your ingenuity to attach the various wires to the center plastic block for both insulation and strain relief. The coaxial cable, which may be any length, is a 72-ohm type sold for tv (RG-59/U).

You may observe some directionality in this antenna array. As in all half-wave dipoles, the most sensitive pickup occurs broadside to the elements. For example, if the antenna runs north and south, it works best on signals arriving from the east or west.

9

TV Antenna for Shortwave

If you want to alternate your tv viewing with your shortwave listening, you can try this trick. It enables you to quickly switch the tv antenna to the shortwave receiver. Although not intended to replace a conventional shortwave antenna, this setup will provide improved reception on some shortwave bands. The only additional part needed is a knife switch of the double-pole, double-throw type widely sold in electronic supply outlets.

The switch provides a fast, convenient method of connecting your tv antenna to your shortwave receiver. It does this by converting the lead-in line to a long-wire antenna. The tv antenna itself is ignored since its elements are so physically short they barely respond to shortwave signals. It is the long run of twinlead that provides the actual signal pickup. When the knife switch is in the "SW" position, the two wires of the twinlead are shorted together so they electrically form a single wire. Note the jumper wire added to the two right-hand terminals of the knife switch, and a length of wire to the antenna terminal of the shortwave receiver (Fig. 1-9). Since the system is not simultaneous, you can only have reception on either the tv or shortwave receiver, depending on which way the switch is thrown.

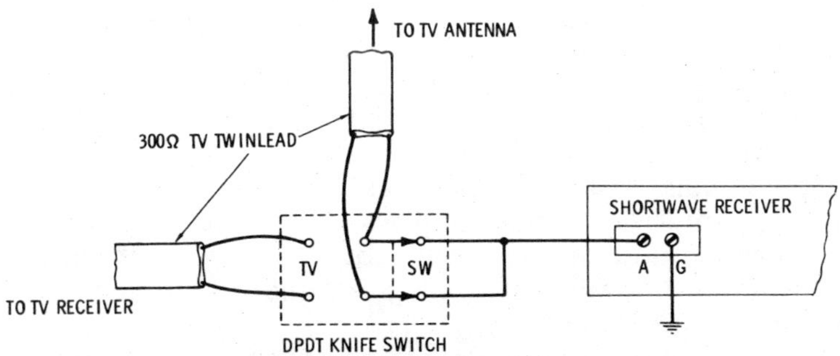

Fig. 1-9. Switching tv antenna to shortwave receiver.

10

"Super" Antenna

What's the best shortwave antenna? It's generally considered to be the *rhombic,* named after its diamond-shaped configuration. For the amount of wire used in its construction, and the range of frequencies it covers, the rhombic picks up more signal than just about any other antenna design. It is widely used for intercontinental communications by commercial and military stations. The antenna is highly directional and operates with a very low angle of radiation, which favors signals arriving from great distances.

The rhombic is also impractical for most shortwave listeners. It requires almost 1000 feet of wire, needs a large plot of real estate, and calls for four support points. But if space and cost are no object, then a rhombic will certainly improve your shortwave reception. Let us consider the rhombic in some detail, since it can be erected with several cost-cutting steps. One fixed parameter, however, is the total ground area; the lot must be about 450 feet long by 180 feet wide. Since the rhombic is directional, this also affects its orientation on the lot. Incidentally, the lot is still usable for other purposes. The antenna simply runs around the perimeter of the lot and is elevated from the ground.

As shown in the top view of Fig. 1-10, there are four 230-foot legs. It is assumed that the house will lie to the left in Fig. 1-10, where the lead-in wire appears. This end of the antenna is most conveniently attached to the house for support. At the opposite end, there is a terminating resistor where the distant legs merge. A support point is needed here, as well as at the two corners near the center. The height of the supports should be about 50 feet above ground for best results, but a somewhat lower height is possible with some sacrifice in performance on the low shortwave bands (about 6 MHz).

If the lot has trees, there is a good chance you can use them for supports and greatly simplify the installation. Start by cutting the four 230-foot legs and laying them out on the ground according to the diagram in Fig. 1-10. The various supports do not have to lie exactly at the antenna corners. It is possible to run guy wires or ropes to nearby trees and maintain the corners at the correct position. A helpful technique is to first draw the lot on paper, showing the possible supports (such as trees) and

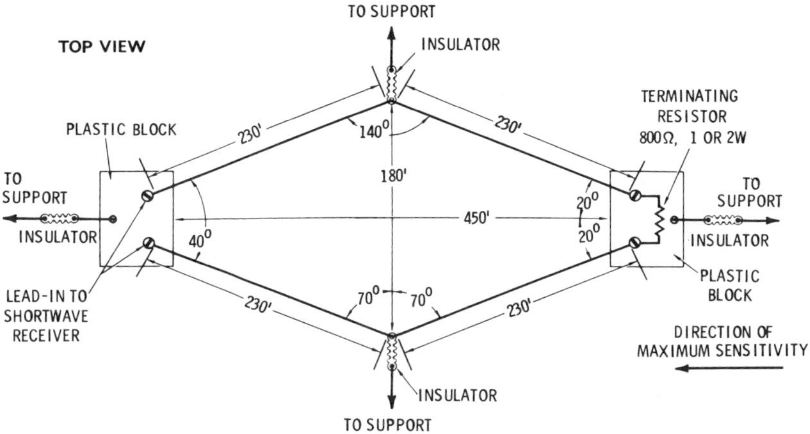

Fig. 1-10. Rhombic antenna.

juggling the antenna layout until you get the best fit and orientation. Use a scale of one inch per 60 feet, and a regular sheet of typing paper will accommodate your sketch.

Once the antenna is in place on the ground, consider further construction details. Use a plastic block several inches square (or some other insulator) and drill it to receive nuts and bolts to hold the lead-in and antenna wires where they merge near the house. The lead-in wire may be 300-ohm tv twinlead or, even better, open-wire ("ladder") line sold in electronic parts stores.

A second insulating block is located at the far end, where the terminating resistor is installed. This is a carbon resistor of about 800 ohms, 1 or 2 watts, which is soldered across the two legs. The block, with suitable hardware, supports these items and also allows you to fasten an insulator, as shown.

The antenna will be highly directional (most sensitive) to signals arriving from the direction of the terminating resistor, as indicated by the arrow in Fig. 1-10. If you wish to sacrifice some gain, but operate the antenna in a bi-directional fashion, remove the terminating resistor. Now the rhombic is equally sensitive to signals arriving from the left, as well as from the right.

SECTION 2

Receivers

11

Mechanical Bandspread

One of the distinguishing features of any shortwave receiver is *bandspread.* This refers to the tuning rate, or ability of the receiver to space signals along the dial. In a low-cost receiver, for example, one complete rotation of a bandspread dial may tune 500 kHz on the dial. In more expensive receivers, one complete rotation may tune only 25 kHz. As bandspread increases, it is easier to separate and select stations.

There are two methods used by circuit designers to improve bandspread: electrical and mechanical. The first may not be practical in your receiver since it requires extensive modification to internal tuning circuits. The second method, however, can be added to almost any receiver except some small portables. This method is the installation of a specially-designed *vernier* dial which gears down the bandspread tuning rate of the receiver. The regular bandspread tuning knob is removed and the new vernier dial installed in its place. These dials are widely available from electronic parts supply outlets and catalogs. Inside the device is a planetary drive which reduces the tuning rate by a factor of about 8 to 1.

A typical vernier dial is shown in Fig. 2-1A. It could replace the bandspread dial on the shortwave receiver shown in the photograph. The receiver dial has a total travel of one-half turn. After the vernier is installed, the half turn is converted to four full turns of bandspread. Tuning "feel" is also improved since the drive operates with smoother action than does the usual bandspread capacitor.

(A) Front view showing receiver bandspread.

(B) Rear view showing method of mounting.

Fig. 2-1. A typical vernier tuning dial used for increasing mechanical bandspread.

There are a few considerations in adding mechanical bandspread. First, you must have room on the receiver panel. The dial shown here is a 2-inch diameter type, but these are also sold in 1½- and 3-inch versions. Check to determine which one will fit your receiver. Mounting of the unit is done through a ¼-inch coupler at the rear (Fig. 2-1B) which slides over the bandspread shaft. Three mounting screws around the edge of the vernier fasten to the front panel of the receiver. A problem you might encounter is getting the vernier to lie flush against the panel. You might have to move the bandspread tuning capacitor back slightly from the front panel with spacers or other hardware. A few moments' study of the problem should suggest the best mounting scheme.

Another important point to remember is that adding mechanical bandspread to improve tuning will not overcome a common problem that many inexpensive receivers suffer from: poor *selectivity* in their internal tuning circuits. These receivers tune so broadly that more than one station is admitted into the receiver at once. No degree of mechanical bandspreading can cure this problem. You should consider the vernier dial only as a method of mechanically decreasing the tuning speed to a more comfortable rate.

12

Antenna Trimmer

This front-panel control (Fig. 2-2A) is often missing from low-cost receivers. If your receiver does not have this trimmer, it is worthwhile to consider adding one. The expense is small, and you might obtain an important increase in receiver sensitivity on certain frequencies. The trimmer is a variable capacitor (Fig. 2-2B) with a rating of approximately 50 pF. It is connected to the input of the rf amplifier stage, as shown in the simplified schematic of Fig. 2-2C. If the receiver has no rf amplifier, the capacitor is connected to the same point in the next stage (the mixer or converter).

(A) Front panel view.

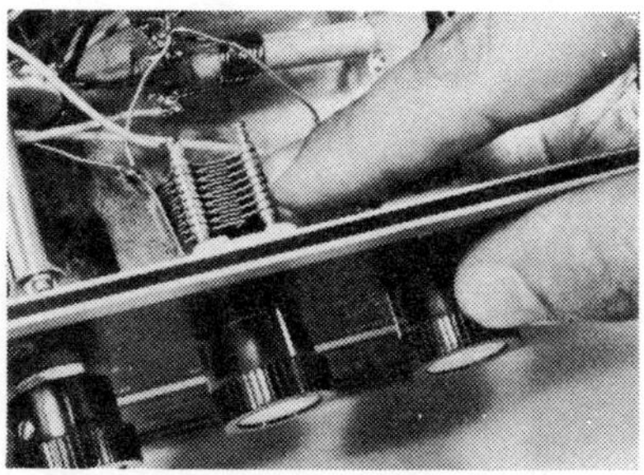

(B) Under chassis view.

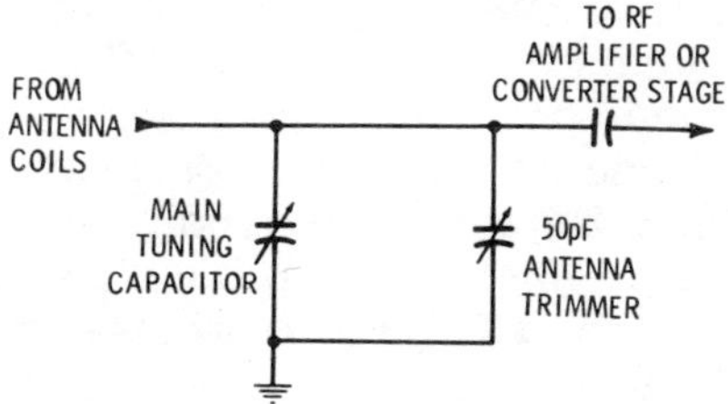

(C) Simplified schematic.

Fig. 2-2. Antenna trimmer.

The job of the trimmer is to compensate for certain variations that cannot be predicted when the receiver is aligned at the factory. For example, your antenna may slightly detune the antenna coils in the receiver. With an antenna trimmer, however, just a slight adjustment of this control can restore the circuit to highest efficiency. The control is used each time the receiver is tuned to a new frequency.

A trimmer also helps to correct a condition present in virtually all receivers. This condition is known as "tracking error," where the two or three variable tuning circuits in the receiver fail to keep in perfect step with each other as they are tuned across a band. The trimmer allows you to compensate for this tracking error in the antenna circuit.

As shown in Fig. 2-2B, the trimmer is a small variable capacitor that has a ¼-inch shaft and a threaded bushing. The bushing fits into a hole that you drill in the receiver panel. The leads to the trimmer should be kept as short as possible by mounting the unit close to the main tuning capacitor assembly. The trimmer is connected across the antenna section of the main tuning capacitor. If the frame of the main tuning capacitor is insulated from the chassis, you must also insulate the trimmer with a suitable insulating bushing. In most cases, however, the frame of the trimmer is connected to the chassis for an electrical ground.

13

RF Gain Control

During the reception of very strong signals, it is possible to overload the early stages in the shortwave receiver. When this happens, the sound becomes garbled and difficult to understand, or you may hear strange buzzes as you tune across the dial.

Modern communications receivers all include an automatic gain control (agc) circuit that automatically reduces strong signals to a level sufficient to prevent overload under normal conditions. At times, however, very powerful signals may be picked up that will override the agc circuit and still cause overload. During these conditions, it is an advantage to have an rf gain control in the rf amplifier circuit so that you can manually reduce extra strong signals to a safe level. During the reception of normal-level signals and weak signals, this control is adjusted for maximum rf gain.

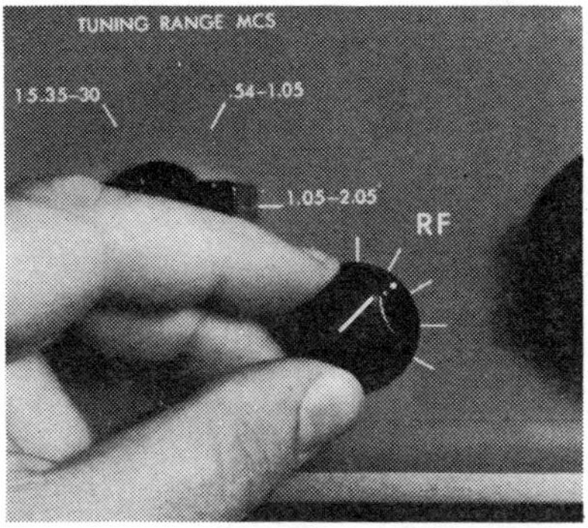

(A) Front panel view.

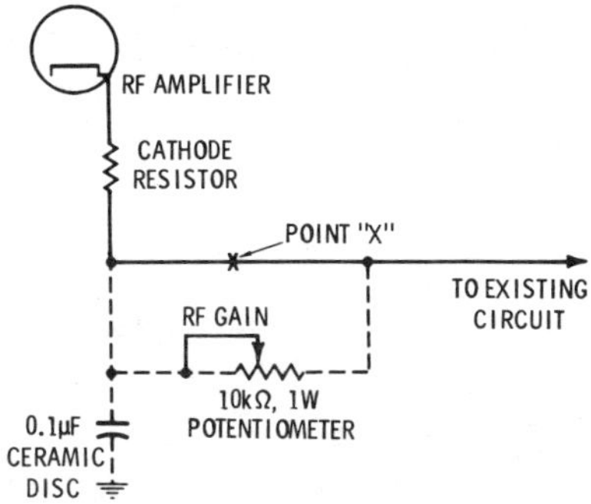

(B) Simplified schematic.

Fig. 2-3. Rf gain control.

Many commercial receivers that have an rf amplifier stage provide an rf gain control as shown in Fig. 2-3A. If your receiver has an rf amplifier stage, but does not have an rf gain control, consider adding one. In a typical tube receiver, this control is a potentiometer connected in the cathode circuit of the rf amplifier tube. The simplified schematic of Fig. 2-3B shows how to connect an rf gain control into an existing circuit. The existing circuit is broken at point "X" and the dotted-in components are connected as shown. The cathode resistor in the existing circuit usually has a value in the neighborhood of 100 ohms. The 0.1-μF capacitor is added to conduct rf currents around the control to ground. After the control has been installed, some experimentation should reveal how to use it to best advantage in a variety of receiving conditions.

14

"S" Meter

Most of the better quality shortwave receivers have an "S" meter mounted on the front panel (Fig. 2-4A). By indicating the relative strength of an incoming signal, the meter provides a clear indication of tuning accuracy. Simply watch the needle for a peak reading as you tune in the station. The meter also indicates signal fading by swings of the needle. Slight fading may not be audible to the ear, but can easily be seen on an "S" meter.

An "S" meter is also a measuring rod of the efficiency of your antenna system. When you make an improvement in your

(A) "S" meter on commercial receiver.

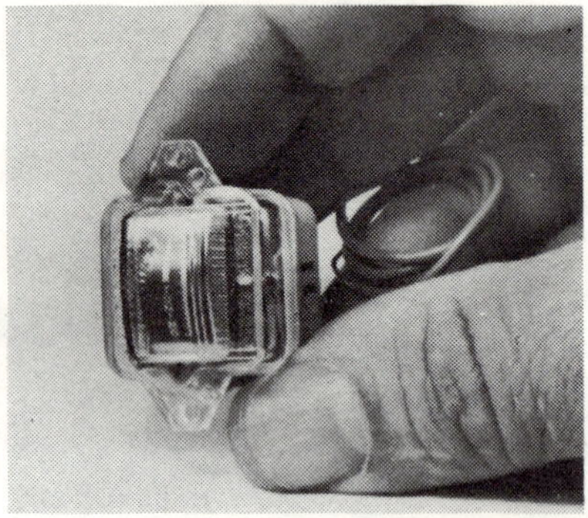

(B) An inexpensive hobbyist-type "S" meter.

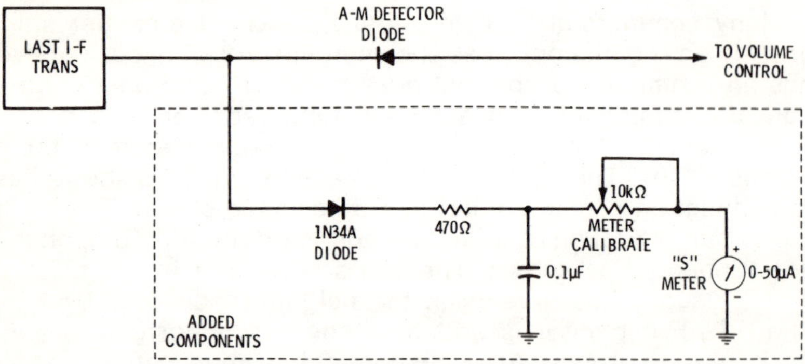

(C) Schematic of "S" meter circuitry.

Fig. 2-4. "S" meter.

antenna system, or move it to a new location, the comparative results will appear on the "S" meter. The meter also provides an indication of the amount of noise or interference being picked up. An "S" meter indicates signal levels that are independent of the volume-control setting.

If your receiver does not have an "S" meter, one can be added without much difficulty. Be careful of the type meter you use, however. Do not attempt to install any of the meters listed in electronic parts catalogs as replacements for other receivers. These receivers already have the circuits to drive a meter. Such meters are usually rated at 0-1 milliamperes dc and are not sensitive enough to operate in the system described here. This system uses a very sensitive meter movement rated at 0-50 microamperes dc. The meter you choose can be an inexpensive unit like the one shown in Fig. 2-4B. These are widely available as hobbyist components in electronic supply stores or catalogs. Just be certain it has the 0-50 microampere rating. Most meters of this sensitivity are rather expensive for this application.

With just a few other components, the meter movement can be connected directly to a point in the intermediate-frequency (i-f) section of the receiver, and operates from the low-level signals at this point. The i-f signal at this point varies directly with the strength of the incoming rf signal at the antenna.

As shown in the schematic of Fig. 2-4C, the i-f signal is taken off between the last i-f transformer and the a-m detector, and applied to a 1N34A diode. (The a-m detector may be either a semiconductor diode or a diode tube.) This is usually an easy point to locate: find the volume control, then trace the wire that leads to the a-m detector diode. The detector will usually be connected to one terminal at the bottom of the last i-f transformer.

The operation of the circuit is straightforward. The 1N34A diode receives a small sampling of i-f signal and converts it to pulsating dc. Since the "S" meter must operate on smooth dc, the 470-ohm resistor and the 0.1-μF capacitor serve as a filtering network. The result is a dc voltage that varies with the strength of the incoming signal. This voltage is then applied through a calibrating potentiometer to the 0-50 microampere meter. These components are mounted inside the receiver, with the potentiometer shaft protruding through a hole in the chassis so you can calibrate the meter.

Since the "S" meter indicates only relative signal strength based on an arbitrary scale of "S" units ranging from "S1" for very weak signals to "S9" for strong signals, the calibration adjustment is approximate and is done by ear. Tune in a good strong signal that just quiets the background atmospheric noise, then adjust the calibrating potentiometer until the needle reads about two-thirds up the scale. This is roughly equivalent to a reading of "S9," and signals weaker than this will be mixed with natural and man-made interference. If the meter you use is already calibrated in "S" units, simply adjust the control for a reading of "9" with a strong signal tuned in.

15

Selecting Headphones

Old timers called them "cans," but today they are known as earphones, headphones, headsets, or just plain "phones" (Fig.

2-5). Regardless of what you call them, headphones are an item of inestimable value for shortwave listening, and are not usually supplied as standard equipment. Headphones have many advantages: You can listen at times when the rest of the family is asleep; you can listen in a room with another person without disturbing him; you can set the volume at whatever level you desire; and, by excluding room noises and acoustics, it is sometimes possible to understand weak signals that are not intelligible through the speaker. (If your set has no headphone jack, you can add one as described in Item 16.) The receiver instruction manual usually gives the electrical rating for phones that may be used with your set (usually a dynamic or magnetic type with an impedance of about 50 to 2000 ohms). There are many models to choose from, so here are some factors to consider before making a purchase.

The biggest drawback to headphone listening is wearing comfort. Phones of poor design become uncomfortable, or even painful to wear after several minutes. The reason is not a matter of weight (even massive phones can be quite comfortable) but how the pressure is distributed. There should be sufficient padding to distribute the pressure as evenly as possible around the ear. The headband must be easily adjustable to ensure that the pads are properly positioned.

There should be a good air seal between the phones and your ear. Otherwise, too much sound is lost or excessive room noise enters the ear. (You may, however, wish to allow some sound to enter your ear from the outside so that you can hear the telephone ringing or someone knocking at the door.) If the seal is tight, then make sure that too much heat does not build up around the ears after wearing the phones for a while. A material that is used to reduce both heating and pressure is the foam cushion. If you can try headphones for more than a minute or two, each of these factors should become apparent. One final consideration is to be certain that the plug on the headphone cord is the right size for the phone jack on your receiver.

16

Adding a Headphone Jack

If your shortwave receiver is not equipped with a headphone jack, here are some general instructions on how to install one. First, find the two leads from the secondary of the audio output transformer. One lead should go to chassis ground, the other

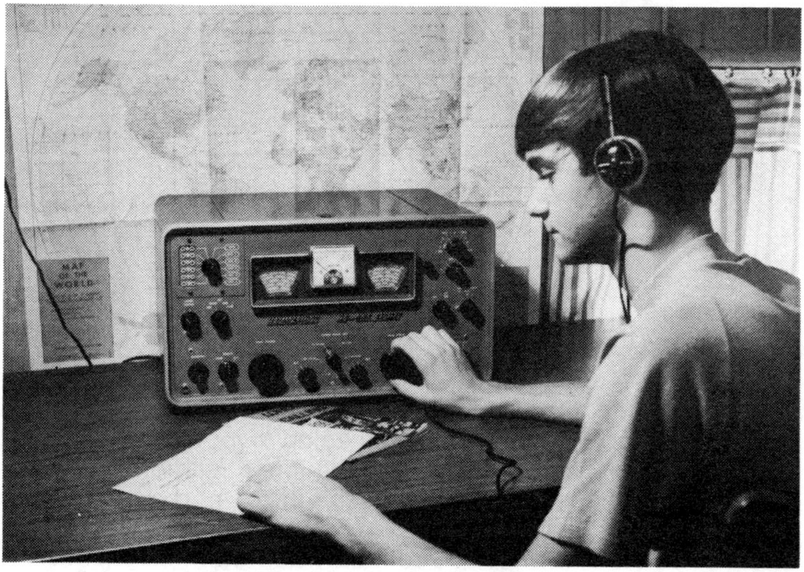

Fig. 2-5. Headphone listening.

to a speaker terminal. Cut the speaker lead at point "X" (see Fig. 2-6) and extend the ends to a "closed-circuit" type phone jack mounted somewhere on the receiver chassis. Next, add a resistor (R), the value of which depends on the speaker impedance. Speakers used in tube receivers normally have an impedance of about 4 ohms, whereas those used in solid-state receivers are about 10 ohms or more. The resistor should have a value of approximately ten times the impedance of the speaker. In tube receivers, this would be about 47 ohms; in transistor receivers, about 100 ohms. In each case, use a 1-watt type so that the resistor can absorb power from the audio amplifier during headphone listening. This prevents overdriving the headphones since they do not require as much power as the speaker does. (When the headphone plug is inserted into the phone jack, the speaker is automatically disconnected.)

The choice of headphones can include any of the general-purpose, magnetic types that have impedances ranging from about 50 to 2000 ohms. If you use low-impedance, hi-fi phones (4 to 16 ohms), the volume may be uncomfortably loud. This can be corrected by first adjusting the receiver volume control for normal listening level from the speaker. Then insert a resistor in series with one side of the headphone circuit to reduce the volume to a comfortable level. The resistance value will vary, so this must be chosen experimentally. Start with a value of several

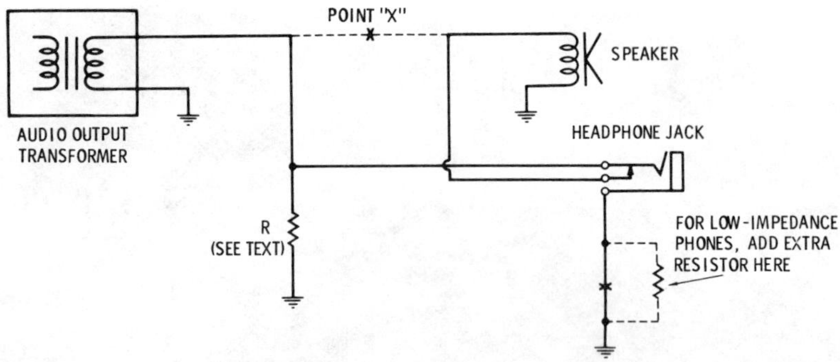

Fig. 2-6. Adding a headphone jack.

thousand ohms, then reduce the value until the volume in the headphones is at a comfortable listening level with the receiver volume control set to the point mentioned earlier. The extra resistor must be inserted in the headphone part of the circuit and not in the speaker circuit, as shown in Fig. 2-6, so that the speaker volume will not be reduced when the headphones are unplugged.

The circuit shown in Fig. 2-6 is not for receivers of the ac-dc "hot-chassis" type. Adding a headphone jack to ac-dc receivers (which have no power transformer) requires special insulation of the jack from the chassis in order to prevent a possible shock hazard.

17

Better Speaker Sound

Because of limited cabinet space, the speakers supplied with most shortwave receivers are small and do not produce good quality sound. This reduces the pleasure of listening to musical programs transmitted by overseas broadcasters. The easiest and quickest method for improving the sound is to use an external speaker and cabinet. An 8-inch speaker, for example, will have much better low-frequency response (bass tones) than the 2- or 4-inch units built into many receivers. You could also select a speaker sold for hi-fi, but you might have difficulty driving an acoustic-suspension type. These are low-efficiency speakers and need a considerable amount of power to generate ample sound. The audio section in most shortwave receivers will deliver about 2 or 3 watts in tube sets, and less than 1 watt in solid-state portables.

There is, however, another course of action open to the hobbyist. He can purchase a speaker of reasonably good quality and construct a suitable cabinet at minimum cost. The speaker cabinet should be made from ¾-inch thick plywood panels which are joined by both glue and screws to prevent vibrations. Screws are used every four inches to assure rigid construction. The inside of the cabinet is lined with fiber-glass batting of about 1-inch thickness to serve as acoustic damping. Select a grille cloth of wide mesh, or one that is sold for hi-fi use to prevent loss of high frequencies.

After the speaker is mounted in a suitable cutout hole on the front panel, the cabinet is sealed with a back cover. The result is an "infinite baffle" enclosure, the performance of which should be superior to that of the original speaker. A small baffle will do, but as the dimensions grow larger, so does the improvement in low-frequency response.

The disadvantage of a good speaker system for shortwave listening is that noise and other interference may also be boosted. If the signals are sufficiently strong, however, the improvement in musical quality can be worth the effort. Some improvement on noisy signals is possible if you have some method of tone control, since a reduction in treble response can reduce the high-frequency hisses and static. Most shortwave receivers do not provide this flexibility. If you want to try yet another approach, which offers excellent possibilities for better sound, see Item 18, "Shortwave to Hi-Fi," described next.

18

Shortwave to Hi-Fi

Although shortwave broadcasting is not usually mentioned as a source of hi-fi programming, it has definite possibilities. If you have a hi-fi set, you can use it to improve the quality of sound from your shortwave receiver. In most cases, the interconnections between the shortwave receiver and the hi-fi amplifier are easily made.

The various tone controls in a hi-fi amplifier introduce a versatile form of control over the quality of audio from the shortwave receiver. For example, if your shortwave receiver has no noise limiter, reducing the treble response can simulate the effect. This could minimize bursts of static and ignition interference from automobiles passing nearby. Adjusting the bass control can help reduce the effects of "fluttering," which is a rapid signal fading that occurs during certain atmospheric conditions. Since flutter can occur at rates that match those of bass notes, adjusting the bass control could eliminate the worst effects.

There are two methods for feeding the shortwave signal into a hi-fi amplifier. Many shortwave receivers have a headphone jack, and this can be a connection point. You will need a "patch cord," which is simply a length of shielded audio cable with plugs on both ends; one to match the headphone jack on the shortwave receiver, and the other to match the auxiliary-input jack on the hi-fi amplifier. This method bypasses the speaker in the shortwave receiver.

You can obtain even better audio by picking off the signal at the output of the a-m detector. This method bypasses the audio stages in the shortwave receiver. To make the connection, you will need a length of shielded audio cable with a plug on one end to match the auxiliary input on the hi-fi amplifier. The other end of the cable is connected to the volume control in the shortwave receiver. As shown in Fig. 2-7, the wire to the center terminal of the volume control is removed, and the inner conductor of the audio cable is connected in its place. The cable shield is connected to ground in the receiver, which is one of the outer terminals of the volume control. If you have difficulty identifying the ground terminal, touch the cable shield to either outer terminal on the volume control to determine which produces the best results through the hi-fi system. This connection is not recommended if the shortwave receiver is an ac-dc type unless you use an isolation transformer between the 117-Vac power line and the power input to your receiver, in order to avoid a shock hazard.

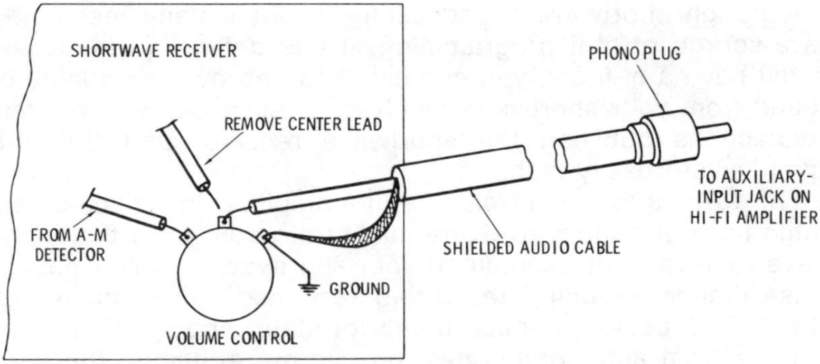

Fig. 2-7. Connecting the shortwave receiver to a hi-fi amplifier.

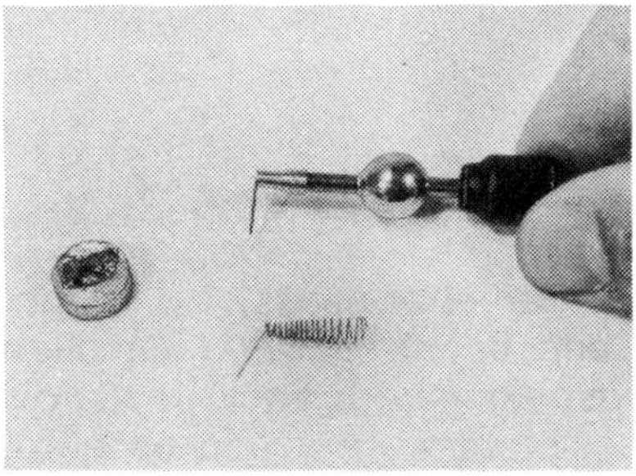

(A) Crystal (left) and "cat's whisker."

(B) Winding the tuning coil.

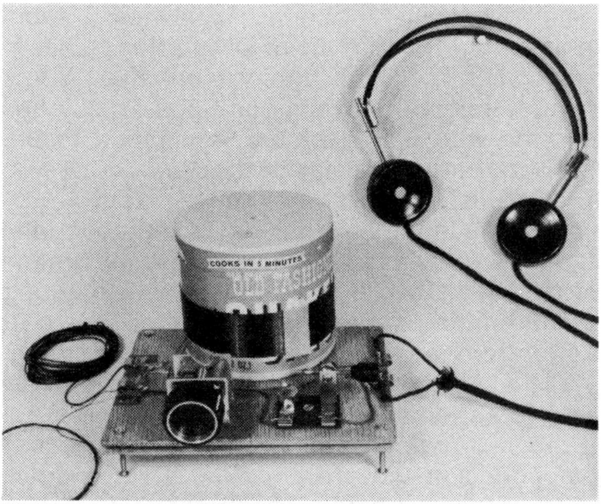

(C) A completed crystal receiver.

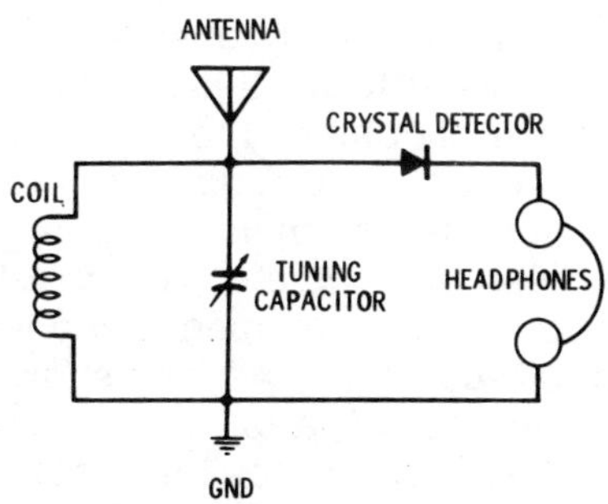

(D) Basic schematic of a crystal receiver.

Fig. 2-8. Making an antique shortwave receiver.

19

Make an Antique Shortwave Receiver

Today's shortwave receiver is a marvel of sensitivity. It captures a few microvolts of signal and boosts it several million times to make it audible in a loudspeaker. But back in radio's earlier days, when tubes and transistors were yet to be born, the crystal set was the only practical method of radio reception. Without benefit of any amplification, these circuits could sometimes capture signals from thousands of miles away.

So intriguing is the challenge of picking up signals from afar with a primitive receiver that it has spawned a subdivision of shortwave listening. It is the art of DXing (long-distance reception) with a crystal set. With barely more than a few turns of wire and simple parts, hobbyists can occasionally hear signals from across the world. One record is claimed to be between Columbus, Ohio, and Makassar in the East Indies—a distance of some 7000 miles!

The components in the circuit today are hardly different from those of the pre-1920 era of radio broadcasting when everyone listened on a "cat's whisker" set. It begins with a long antenna wire, a variable tuning capacitor, and a coil. These items capture the signal from the air and separate it from all the others in the vicinity. The "cat's whisker" refers to a tiny, sharp wire used to feel over the surface of the crystal (see Fig. 2-8A), which is actually a piece of galena (an ore of the metal lead). Natural crystals embedded in the ore serve as a radio detector. This function, done today by a semiconductor or tube diode, separates the audio frequencies from the radio signal. The result is fed to earphones for listening. All the power in the circuit is derived from the tiny energy intercepted by the antenna.

If you want to try this interesting, but demanding, branch of shortwave listening, the parts are available in the catalogs of several major electronics distributors. A good place to search for components, plans, and instructions is in the classified advertisements of electronic hobby magazines. They usually list good sources of specialized information about long-distance crystal receivers. The set shown in Fig. 2-8 was built from such plans and parts. Do not try for distance with one of the miniature diode radios that "needs no battery." These sets may work on strong local broadcast stations, but are not efficient enough for long-distance reception.

SECTION 3

Accessories

20

Quick Time Converter

Let's see; when it's 5 p.m. in Chicago, it's 8 a.m. in Hong Kong, 1 p.m. in Honolulu, and midnight in Paris. These calculations aren't the work of a mathematical wizard, but were quickly dialed on a wheel like the one shown in Fig. 3-1. These inexpensive cardboard computers show at a glance the equivalent time anywhere in the world. They're also useful for converting GMT

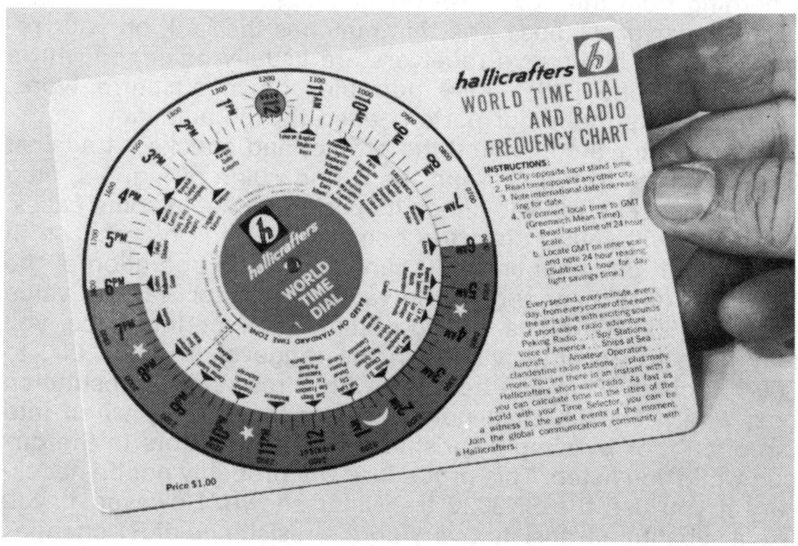

Fig. 3-1. Quick time converter.

(Greenwich Mean Time) into local time. Since many shortwave schedules are announced or printed in GMT, a time-conversion computer is handy for this reason, too.

21

Headphone Amplifier

Most receivers provide plenty of audio gain for headphone listening. Some small portables, however, may prove deficient and fail to drive headsets to a comfortable volume level during weak-signal reception. Here is a one-transistor amplifier that can make up for the loss. It is easily constructed and plugs into the headphone jack on the receiver. Another jack is mounted on the amplifier to receive the plug from the headphones. The circuit is designed to operate with magnetic phones of a few hundred to several thousand ohms impedance. Do not use hi-fi phones of about 8-ohms impedance because these are designed to connect directly to the low-impedance output of a hi-fi audio amplifier.

As shown in the photographs of Fig. 3-2A, B, and C, the circuit is built into a small plastic case. (If you prefer, you can also construct it on a small piece of perforated circuit board.) A standard phone plug is shown attached to the shielded cable emerging from the box; however, this item will vary according to the receiver. Choose one that matches the jack on your receiver. Large shortwave receivers are usually equipped with a standard phone jack, while portables have miniature types. Table 3-1 lists the major parts needed for the amplifier.

After you have completed the project and checked it against the schematic of Fig. 3-3 and the wiring guide of Fig. 3-4, plug it in to your receiver and check its performance. In some cases, the audio will sound distorted because of small differences in receivers. If this happens, you can change the operation of the transistor (Q1) by changing the value of resistor R2. The value shown is 330,000 ohms, but to obtain better performance, you may have to try other values in the range of from 100,000 to 500,000 ohms. This can be done with a resistance substitution box, with a 500,000-ohm potentiometer temporarily wired into the circuit, or by simply substituting other resistors in the circuit while you listen. This procedure will probably not be necessary if you use the SK3020 transistor shown. However, if you use a substitute transistor, varying the resistance of R1 changes the operating bias to match the particular transistor.

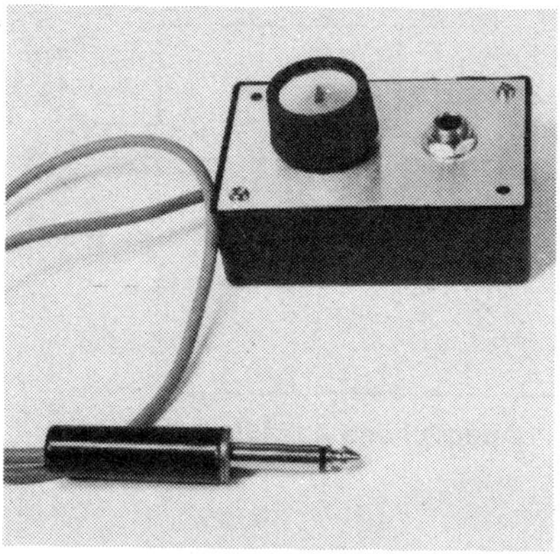

(A) Completed unit.

(B) Inside view showing wiring details.

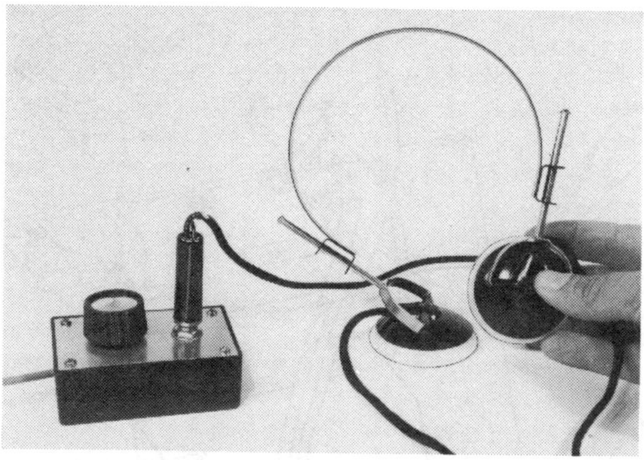

(C) Completed unit with headphones attached.

Fig. 3-2. Headphone amplifier.

Table 3-1. Parts List for Headphone Amplifier

Item	Description
C1	Capacitor, 0.1 μF, paper or mylar.
J1	Phone jack (to match plug on headphones).
M1	Battery, 9-volt (transistor-radio type).
PL1	Phone plug (to match jack on receiver).
Q1	Transistor, npn (RCA SK3020 or equiv.).
R1	Potentiometer, 5000 Ω miniature, with spst switch.
R2	Resistor, 330 kΩ, ½ W.
R3	Resistor, 330 Ω, ½ W.
SW1	Switch, spst (part of R1).
	Plastic case, 3¼″ × 2¼″ × 1¼″.
	Terminal strip, 5-lug.
	Battery clip, for M1.
	Knob, for R1.

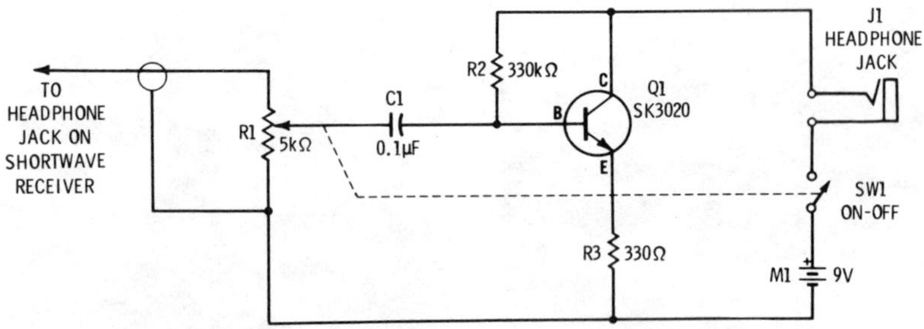

Fig. 3-3. Schematic diagram of headphone amplifier.

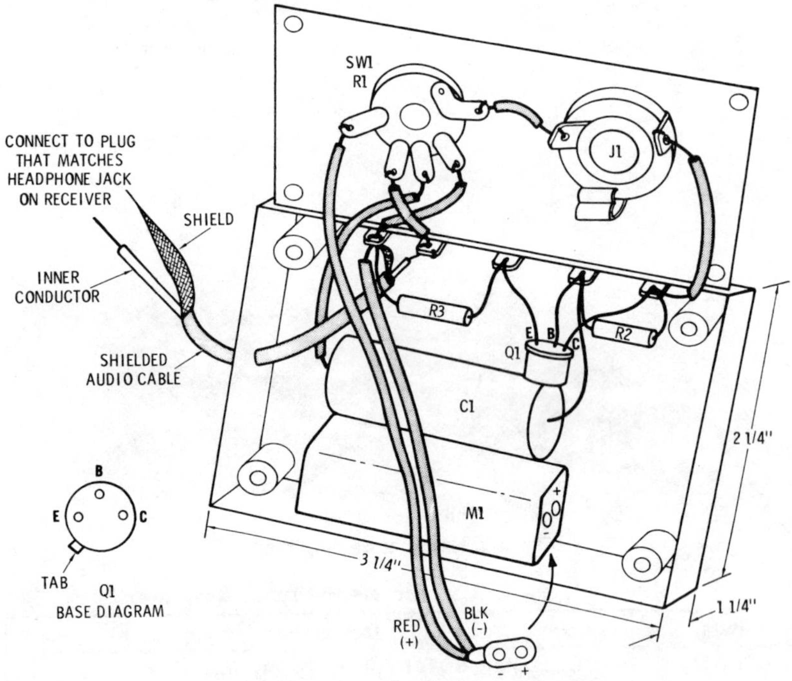

Fig. 3-4. Pictorial wiring guide for Fig. 3-3.

22

World Maps

One enjoyable sideline of shortwave listening is to tune in a signal from a distant part of the world and then locate on a map

Fig. 3-5. Wall map.

where it's coming from. There are not many better ways to learn geography, and it does not take the shortwave listener long to memorize the location of most countries around the globe.

A wall map should be within sight of the listening area (Fig. 3-5). This type of map is handy for finding different countries, but it should not be used for measuring distances. If you place a straightedge on this type of map between your town and some distant country in an attempt to measure the actual mileage of a signal received, the result will contain two errors.

The first error is a minor one. Shortwave signals do not travel on the surface of the earth, but skip back and forth between the earth and the ionosphere some 200 miles above the earth. These upward and downward slanting paths add to the total mileage a signal travels before it reaches your home.

The second error is a major one, and not only affects your measurement of the distance a signal travels, but also the direction from which it arrives. This error is caused by the type of projection used on most flat, wall-type world maps. The difficulty lies in the fact that the earth is a globe, and any projection of the earth's surface onto a flat piece of paper injects distortion (more or less, according to the type of projection) of both distance and direction into the picture, or map. The most ideal map to work with, of course, is the global type, but most of these are rather expensive. Measuring the shortest distance between two points on a globe, though, is easy and is the most accurate way. Simply lay a cloth tape measure, or piece of string, across the globe until you find the shortest distance

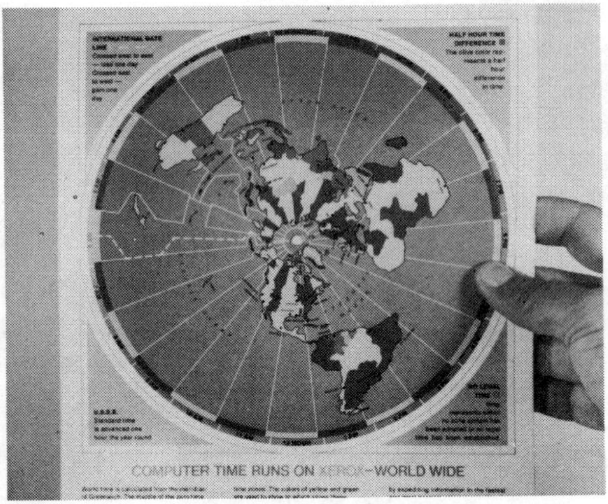

Fig. 3-6. Great-circle map.

between any two geographical points. This measurement is known as the "great-circle" path; a term well known to airline pilots and to anyone who has done much overseas flying. A plane flying the great-circle route between California and the Orient, for example, travels a northerly course which takes it near the lower portion of Alaska and over the Aleutian Islands. This distance is considerably shorter than a path directly across the Pacific Ocean—a fact that is not apparent on most flat-projection maps, but one that is most obvious on a globe.

There is one type of flat map projection, however, which closely approximates the actual distances and directions between points on the globe. This type is properly termed "azimuthal equidistant projection," but such maps are more commonly known as "great-circle" maps (see Fig. 3-6). Note that this map is centered on the North Pole, but the center point can be anywhere. A straightedge placed between two geographic points on this chart is almost as accurate an indicator of true mileage and direction as the string used on the globe.

Radio signals follow great-circle paths around the earth naturally; thus, great-circle maps are very accurate indicators of radio mileage and direction. Actually, there are two great-circle paths for radio signals since the signals may be transmitted in opposite directions from an antenna simultaneously. Thus, the same signals can arrive at your antenna from opposite directions. In most instances, however, you will only be receiving the signal that arrives over the shortest path.

One practical application of the great-circle map is the orienting of a directional receiving antenna. When an antenna is cut to a specific frequency, as described elsewhere, it is directional, meaning that it displays highest sensitivity from certain directions only. The most common example of a directional antenna is the half-wave dipole. Since the dipole is most sensitive to signals arriving broadside to its length, it should be oriented accordingly. The great-circle map helps you to choose the best direction in which to orient your antenna to receive signals from specific points on the globe.

23

Surge Protection

Most of the wear on electronic components occurs when a "cold" circuit is first turned on. Two conditions generally account for this: overvoltage and current surges. Overvoltage occurs because the power-supply voltage tends to rise beyond normal limits until the complete circuit is operational. (Tubes, for example, take time to warm up.) Current surges occur because large electrolytic capacitors may be in a discharged condition and absorb a large inrush of current when the power is applied. Also, in tube circuits, surge currents occur because the tube filaments have less resistance when cold than when warmed up. This means that when the receiver is first turned on, far greater than normal currents flow through the tube filaments. Turning a tube set on and off several times a day shortens tube life because of these continual surges of current through cold filaments. Other components, such as resistors, also suffer thermal damage from repetitive current surges, and this damage is accumulative in its effects. For these reasons, you should turn your equipment on and off as few times as possible. If you are not going to listen to your receiver for a short time, the circuit components will probably last longer if you leave the set on while you are away rather than switching it on and off. The fraction of a penny you save in your electricity bill by turning your set off could be far less than what you save in circuit troubles by leaving the set turned on.

Some ham-radio operators who have tube-type receivers *never* turn them off. Although this equipment may even last longer than equipment that is regularly switched on and off, the ham's main reason for doing this is not protection against surges. The reason is to achieve better stability in the oscillator stage of the receiver. This keeps tuning and dial calibration very accurate by reducing oscillator drift. These are extremely important factors to consider when operating on exact frequencies and while working in the critical single-sideband mode of communication. The shortwave listener may not want to go to such extremes, but in the case of a tube-type receiver, it is advisable to at least turn on the set about a half hour prior to listening, in order to allow the circuits time to stabilize. This is less important in solid-state equipment because of negligible heating.

Another method for minimizing warm-up surge is to use a thermostatic type of surge protector. This is a resistance element that is thermostatically controlled by a bimetal strip. When the circuit is cold, the points on this unit are open. When the receiver is first turned on, this places a resistor in series with one side of the ac line to limit the surge current. The current eventually heats the resistor sufficiently to cause the bimetal strip to close and switch the resistor out of the circuit. The bimetal strip stays heated enough to keep the points closed because of the inherent resistance of the strip itself.

These devices are known by a variety of trade names including *surgistor, tube saver,* or *tube guard,* and can be purchased either for installation inside the chassis (in one side of the ac line) or as separate plug-in devices that are plugged into an ac wall outlet.

24

Crystal Calibrator

Unless you own very expensive equipment, you know that a receiver tuning dial is only approximate in its accuracy. Even expensive units require some help from a built-in crystal calibrator that serves as a frequency reference. A crystal generates extremely precise signals for setting the receiver dial to an exact frequency. Thus, if a station is operating on a frequency of 9520 kHz, for example, you can tune to it with far greater confidence than by following the dial markings. You can easily assemble a small calibrator that will provide you with all the benefits of the calibrators built into more expensive receivers. Photographs of the completed unit are shown in Fig. 3-7A, B, and C.

This is a 2-transistor circuit (Fig. 3-8) which operates as a flip-flop oscillator, meaning that each transistor switches the

(A) Top view.

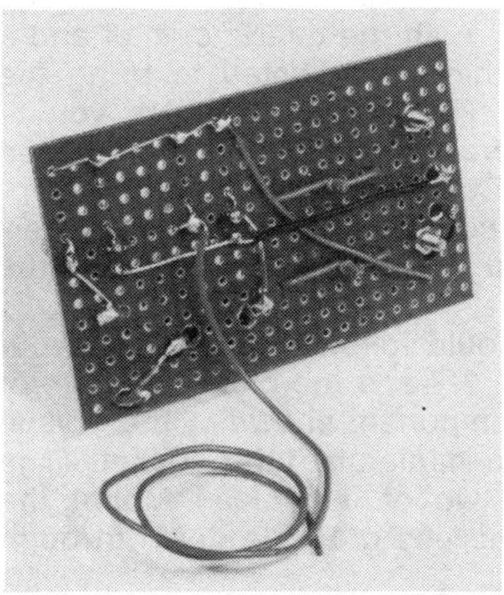

(B) Bottom view.

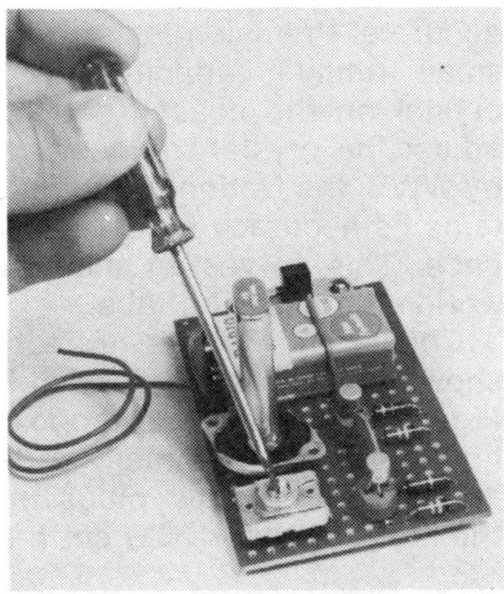

(C) Adjusting crystal trimmer.

Fig. 3-7. Crystal calibrator.

other off and on. Switching speed is controlled by a quartz crystal (X1) cut for 100 kHz. Only when the oscillator is on that frequency will proper feedback action occur from transistor Q2 back to Q1.

Although the fundamental frequency of the oscillator is 100 kHz, it is rich in harmonics that extend all the way up through the shortwave bands. It is possible to hear signals beyond the 200th harmonic, or beyond 20 MHz. This assures dozens of

Table 3-2. Parts List for Crystal Calibrator

Item	Description
C1	Capacitor, .01 μF ceramic disc.
C2	Capacitor, 24-200 pF, mica trimmer (Elmenco Type 425 or equiv.).
M1	Battery, 9-volt (transistor-radio type).
Q1, Q2	Transistors, npn (RCA SK3020 or equiv.).
R1, R3	Resistors, 100 kΩ, ½ W.
R2	Resistor, 560 Ω, ½ W.
R4	Resistor, 1000 Ω, ½ W.
SW1	Switch, slide or toggle, spst.
X1	Crystal, 100 kHz frequency standard.
	Perforated phenolic board, 4″ × 2½″.
	Crystal socket, or 9-pin miniature tube socket.
	Battery clip, for M1.
	Transistor sockets (2), for Q1 and Q2.

lower harmonics sprinkled throughout the lower shortwave bands to serve as a tuning reference. If the 100-kHz fundamental frequency is correct, then all harmonics will also be accurate. Precision is assured by both the quartz crystal and a small trimmer capacitor (C2), which is adjusted to bring the crystal on the exact frequency. The trimmer enables you to adjust the crystal frequency to beat with a signal transmitted by WWV, the National Bureau of Standards station, which maintains its accuracy by oscillators controlled by an atomic reference. This serves as the primary standard for much of the world's engineering labs.

Construction of the project should follow the general layout shown in the photographs of Fig. 3-7 and the pictorial guide of Fig. 3-9. The only unorthodox component shown is the crystal holder, which is actually a 9-pin miniature tube socket. (The crystal conveniently plugs into two of its holes.) Mount the trimmer capacitor (C2) so that its tabs protrude down through

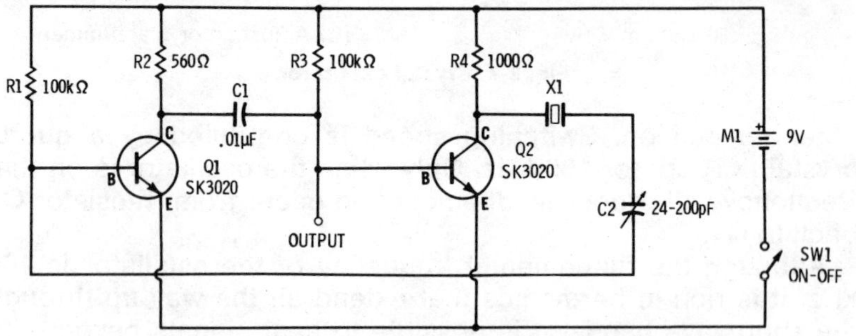

Fig. 3-8. Schematic diagram of crystal calibrator.

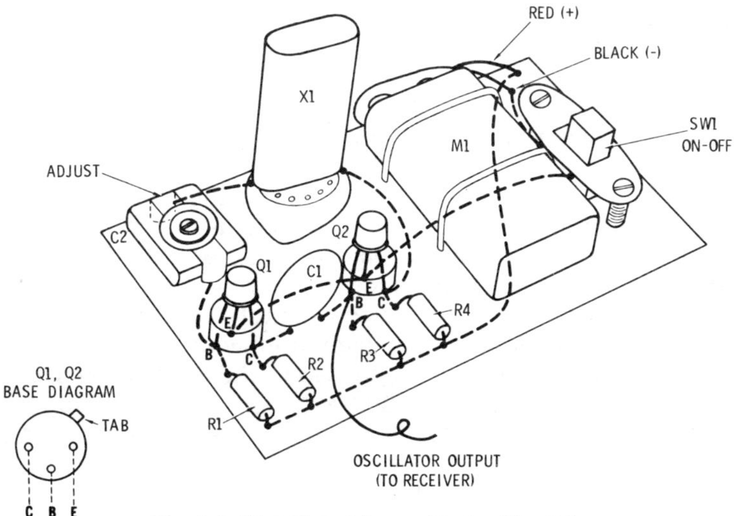

Fig. 3-9. Pictorial wiring guide for Fig. 3-8.

two slots cut in the perforated board. You can install the completed circuit inside the cabinet of your shortwave receiver and bring the on-off switch (SW1) out to the front panel for easy operation. Avoid placing the battery or crystal too close to a hot vacuum tube.

There is just one connection to the shortwave receiver. A length of insulated wire (visible in the photographs of Fig. 3-7) is brought from the output terminal of the oscillator and is wound for about a half-dozen turns around the wire that runs to the antenna terminal of the receiver. This may be done either inside or outside the receiver cabinet. There is no electrical connection since sufficient coupling should exist between the wire turns and the antenna lead to inject the reference signal. If you are using a coaxial lead-in, you should not wrap the oscillator lead around it because of the shield, so attach the output to one antenna terminal through a small capacitor (about 25 pF).

Test the completed project by turning it on and listening to the receiver as you turn the dial. There should be a strong carrier signal that appears at every 100 kHz on the dial. If your receiver has a beat-frequency oscillator (bfo), you will hear the oscillator signal as a series of whistles that go in and out of zero beat as you turn the dial. There will also be other whistles as the bfo beats against stations, but you can verify the oscillator signal by switching the bfo on and off. Next, locate one of the several WWV signals at 2.5, 5, 10, 15, 20, or 25 MHz. These signals are identified by a distinctive "ticking" sound or by voice announcements. Adjust the trimmer capacitor (Fig. 3-7C) until you can exactly zero beat the calibrator with WWV. The tone must go through a null point as you make this adjustment. Do not confuse the zeroing tone with the steady audio tone sometimes transmitted by WWV.

Once this is done, you are ready to use the calibrator. Set up the receiver on the 100-kHz point closest to the frequency you wish to hear. If you wish to tune in a station on 9520 kHz, for example, the nearest calibration point would be 9500 kHz (or 9.5 MHz). Turn on the crystal calibrator and slowly rock the main tuning dial on your receiver until you obtain zero beat with the closest calibrator tone. Now, the bandspread dial is very precisely adjusted to 9500 kHz and you can judge with excellent accuracy how far to move the bandspread pointer to reach 9520 kHz. Once the receiver is calibrated, switch off the oscillator to prevent any interference. You will find that varying the bandspread more than a few hundred kilohertz will probably require another dial calibration, as just described.

25

Build a BFO

The bfo (beat-frequency oscillator) is a standard feature on most of the better quality communications receivers for code (cw) and single-sideband (ssb) reception. It is also helpful in locating extremely weak a-m stations. If your receiver does not have a bfo, one can easily be added and will serve you to good advantage. Without a bfo, code reception is almost impossible because all that can be heard is a sort of "puffing" sound, and this is just barely audible. The bfo provides audible code reception and allows you to participate in code practice sessions conducted over the air by the American Radio Relay League for prospective ham operators. The use of the bfo for single-sideband reception will be discussed later in Section 5.

The basic theory of the bfo is that two signals combined within a detector circuit will mix and produce a third signal, which is the difference in frequency between the original two. (This process also produces a fourth signal, which is the sum of the original two frequencies, but this signal is of no consequence in this discussion.) The difference signal is known as a heterodyne, or "beat-frequency" signal. Even when the two original signals are both beyond the range of human hearing, the heterodyne process can generate an audible tone if the frequency difference between the original two is about 15,000 Hz or less. (This is the upper threshold of hearing for most people.) Thus, the most practical application of the bfo is for creating an audible tone for code (cw) signals. These signals have no audio modulation, and carry intelligence only through the simple on-off keying of the radio carrier.

In almost all radio receivers, the incoming carrier signal bearing code, voice, or other modulation, is converted to a standard intermediate frequency (i-f) of 455 kHz. This is done for several reasons, the most important of which is to make the tuning circuits simpler and more highly selective. The intermediate frequency, however, is still beyond the range of human hearing, so some type of detection process is necessary to remove the modulation and make it audible. The detection process for code reception is done with the bfo. The bfo generates a signal that has a frequency at or near the intermediate frequency of 455 kHz. This is done with a separate oscillator that is built into the receiver and is switched on with a front-panel control (Fig. 3-10). When the i-f signal mixes with the bfo signal, the result is a heterodyne, or audible beat-frequency signal. If the bfo is tuned slightly higher than the i-f signal in frequency to 456 kHz, for example, a difference-frequency signal of 1000 Hz results, and the listener hears the cw code signal at that frequency. (Note that tuning the bfo lower in frequency to 454 kHz, for example, will also produce a 1000-Hz difference-frequency signal). The circuit is adjustable so as to produce any desired tone within the audible range. Most cw operators occasionally vary the frequency of the tone to prevent listening fatigue.

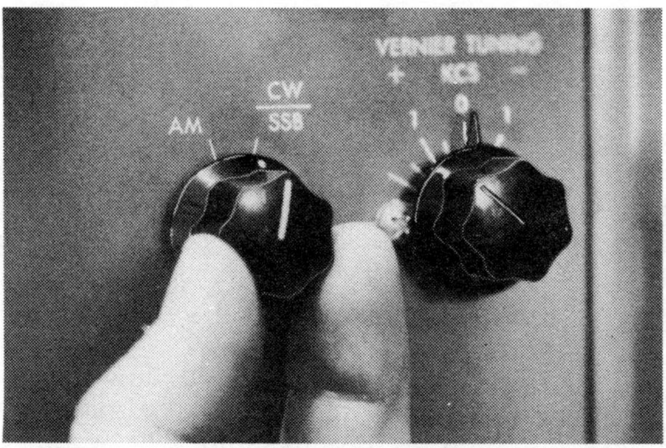

Fig. 3-10. Bfo switch on commercial shortwave receiver.

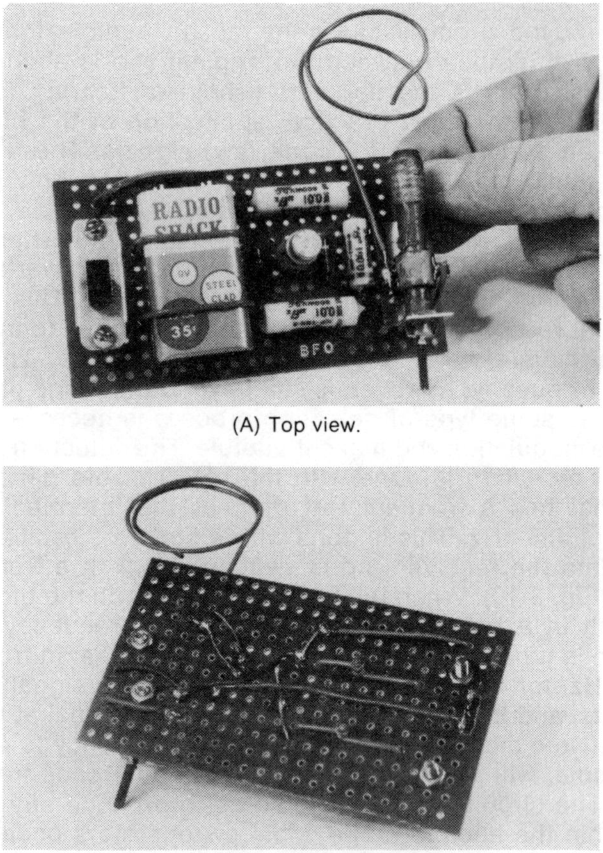

(A) Top view.

(B) Bottom view.

Fig. 3-11. Beat-frequency oscillator.

The practical bfo presented here consists of a one-transistor oscillator circuit which generates a signal in the i-f range of 455 kHz. This unit should radiate sufficient energy to inject a satisfactory signal when placed near the shortwave receiver. As shown in the photographs of Fig. 3-11, the unit is constructed on a small piece of perforated phenolic board. A list of parts for the circuit is given in Table 3-3. A schematic diagram is shown in Fig. 3-12, and a wiring guide is shown in Fig. 3-13. In constructing the circuit, be certain you correctly identify the terminals on coil L1. These must be properly connected in order for the circuit to generate the proper feedback to sustain oscillations. The identity of the terminals can be verified with an ohmmeter—there should be less resistance between terminals 2 and 3 than between terminals 1 and 2.

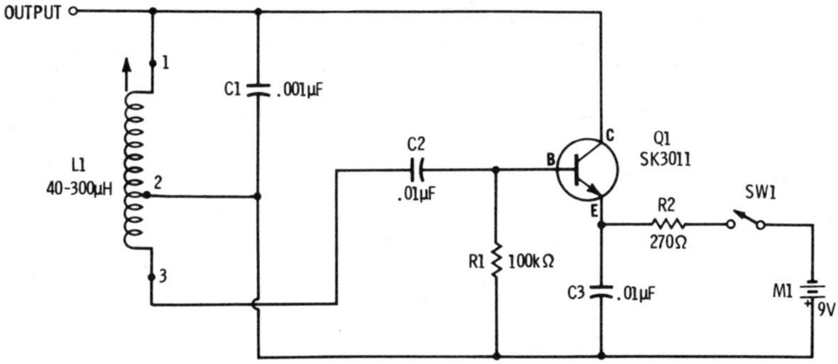

Fig. 3-12. Schematic diagram of beat-frequency oscillator.

Table 3-3. Parts List for Beat-Frequency Oscillator

Item	Description
C1	Capacitor, .001 μF, paper or mylar.
C2, C3	Capacitors, .01 μF, paper or mylar.
L1	Antenna coil, tapped, 40-300 μH (J. W. Miller Type 2002 or equiv.).
M1	Battery, 9-volt (transistor-radio type).
Q1	Transistor, npn (RCA SK3011 or equiv.).
R1	Resistor, 100 kΩ, ½ W.
R2	Resistor, 270 Ω, ½ W.
SW1	Switch, slide, spst.
	Perforated phenolic board, 4″ × 2½″.
	Battery clip, for M1.
	Transistor socket, for Q1.

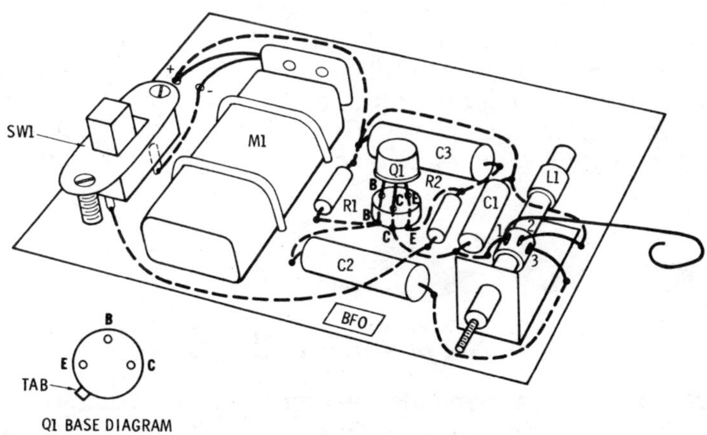

Fig. 3-13. Pictorial wiring guide for Fig. 3-12.

The completed unit is checked by first turning the coil slug until it is approximately centered within the coil winding, then turn on the bfo and the shortwave receiver. Switch the receiver to the standard a-m broadcast band and tune in a local station. Now, tune the slug of coil L1 until you hear a strong beat signal (whistle) fade in and out behind the radio-station signal. Leave the bfo for a moment and tune the radio dial completely across the band (from about 600 to 1600 kHz). If you hear the whistle fade in and out behind *every* station, the bfo is correctly adjusted. If you hear the whistle behind only *one* station, the bfo is tuned to too high a frequency and the slug in L1 must be readjusted to a lower frequency until you find a setting that produces the whistle effect behind every station, including those on the shortwave bands. (One bfo adjustment should take care of all stations, since all incoming signals are converted to 455 kHz). Although the bfo is primarily designed to generate 455 kHz, it will actually cover a range of from about 300 to 1500 kHz, and thus must be tuned to the correct frequency.

As mentioned earlier, another valuable use for a bfo besides code reception is locating weak a-m stations. Turn on the bfo when you suspect a station is tuned in, then listen for the whistle to fade in and out on each side of the station signal. This can often pinpoint a station signal when the audio is too weak to understand because of interference or atmospheric conditions.

On communications receivers that have a built-in bfo, the manufacturer usually also provides a switch to turn off the agc (automatic gain control) circuit. The reason for this is that a combination of a weak station signal and strong bfo signal could cause poor receiver sensitivity. The bfo signal actually looks like a strong station signal to the receiver, and "fools" the agc circuit into reducing the sensitivity of the receiver. This could cause you to miss weak stations when using the bfo. The remedy is to reduce the output of the bfo to a sufficiently low level to prevent such action. This can be done by shortening the bfo output wire, or by moving the bfo farther from the receiver.

26

Aeronautical Charts

World maps, as mentioned earlier in this section, are invaluable for shortwave listening. This can be carried a step further if you become particularly interested in one part of the world.

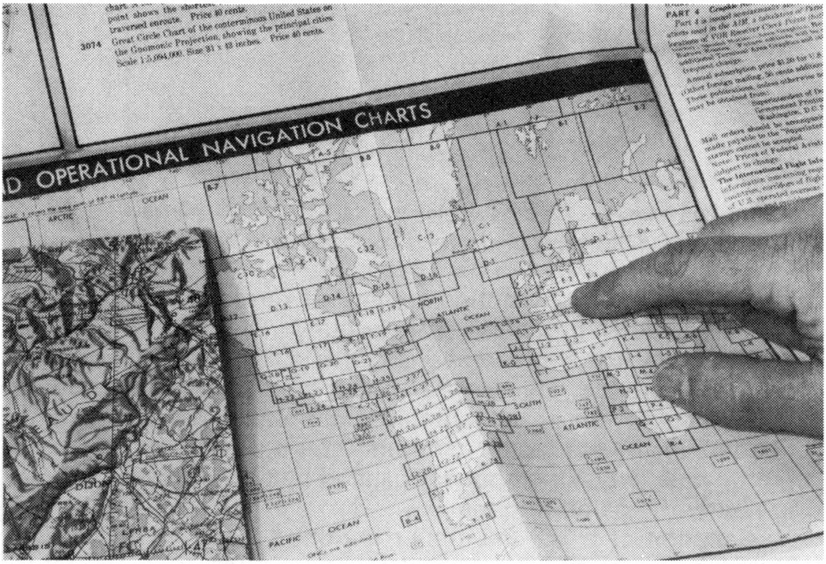

Fig. 3-14. Selecting an Operational Navigation Chart from a chart catalog.

The National Ocean Survey issues excellent charts used by aircraft throughout the world that are not only inexpensive, but contain a huge amount of topographical information. They could prove fascinating to anyone wanting a close, detailed view of a specific region. The scale of many of these charts is 1:1,000,000, and it takes approximately five of them to cover the European continent.

You may choose a chart (they cover about 90% of the earth's surface) by first consulting a chart catalog (see Fig. 3-14). The catalog may be obtained by asking for: "Catalog of Aeronautical Charts and Related Publications" from Distribution Division (C-44), National Ocean Survey, Washington, D.C. 20235. When you receive the catalog, note especially the section on Operational Navigation Charts.

27

"Q" Multiplier

As an add-on accessory, the "Q" multiplier can improve the selectivity of a low-cost receiver. Interference between closely spaced stations can be reduced and improvement often noted through a reduction in background noise. In using the instru-

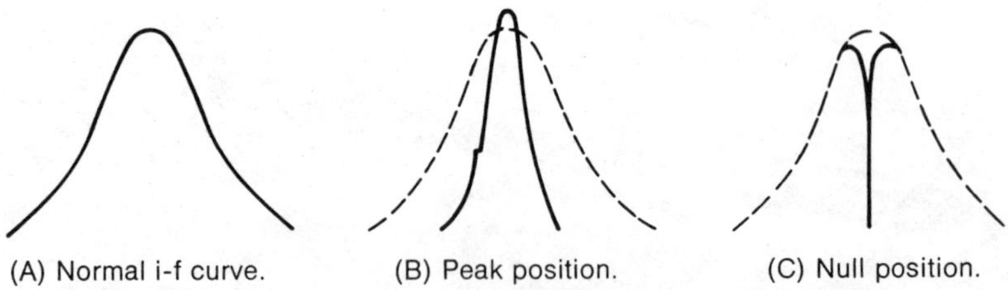

Fig. 3-15. "Q" multiplier curves compared to normal i-f curve.

ment, the listener may notice somewhat reduced audio quality (possibly fewer high tones). This is due to sideband cutting as the circuit narrows the response of the receiver in order to reject interference. This is often offset by an increase in signal intelligibility so it is an acceptable compromise. The "Q" multiplier can be adjusted to an extremely sharp position to eliminate heterodynes (whistling sounds) caused by two stations occupying nearly the same frequency.

Installation of a "Q" multiplier is done through a cable and several power wires into the shortwave receiver. It operates across an i-f transformer to produce extremely high "Q"—a quality which indicates the efficiency of a tuned circuit. A conventional tuned circuit may display a "Q" of 200, while the add-on "Q" multiplier can raise the "Q" to as high as 4000. This degree of sharpness can narrow the response of the receiver or, in reverse, inject a "null" to drop out an interfering signal. A comparison of the response curves is shown in Fig. 3-15. A "Q" multiplier is a worthwhile accessory for the serious listener whose present receiver has neither a crystal filter nor some other internal circuitry (like dual conversion) for obtaining high selectivity. It is generally sold through ham radio distributors and is often available in used condition.

SECTION 4

QSLs, Skeds, and Logs

28

Join a Club

Shortwave listening is greatly enhanced if you can share your DX discoveries or technical tips with others. That's why numerous clubs have sprung up around the world exclusively for the SWL. Their publications are filled with news, rumors, technical ideas, frequencies, schedules, and other items of interest. Some of these publications are simple mimeographed sheets, while others are well-edited bulletins. To join a club and receive its information will cost on the average of from $5 to $15 per year. This often includes airmail delivery from overseas. Here are ten clubs you can contact for more details:

American Shortwave Listeners Club
16182 Ballad Lane
Huntington Beach, Calif. 92647.

Australian Radio DX Club
22 Howard Street
Glen Iris
Victoria 3146, Australia

Danish Shortwave Club International
Tavleager 31, DK-2670-Greve Strand
Denmark

International Radio Club of America
P.O. Box 605
Beaverton, Oreg. 97005

Japanese Shortwave Club
C.P.O. Box 29
Sendai, 980-91 Japan

National Radio Club
P.O. Box 99
Cambridge, Mass. 02138

New Zealand Radio DX League
P.O. Box 1356
Christchurch, New Zealand

Newark News Radio Club
25 Market St.
Newark, N.J. 07101

North American Shortwave Association
P.O. Box 989
Altoona, Pa. 16603

World TV-FM DX Association
P.O. Box 5001
Milwaukee, Wis. 53204

29

Keep a Log

A written record or summary of what you hear on shortwave can be a very enjoyable and practical aspect of shortwave listening. As you look back over its pages, you will relive your most interesting listening. The practical side of logkeeping is that it becomes a reference for preparing signal reports to secure QSL cards and verification reports from distant stations. The log can be kept in an 8½″ × 11″ bound notebook having ruled paper. These are available at nominal expense in most drugstores, discount stores, etc. A spiral-bound type is recommended since it lies flat on the table and takes up the least room.

You will need to rule about ten vertical columns on each page to enter the information about each station. (Some sample logs are provided at the end of the book.) Although you may want to vary the format, you should include the following essential information:

Start with the *date,* then provide a column for the **station's** *call sign* or *name.* Enter the station's *geographic location* in another column. Give the frequency by *band* and *MHz* in the next column, and provide a column for the *time* (in local or GMT). Judge the signal and enter a report, using the SINPO or RST methods described in Item 30, "Composing a Signal Report."

Allow a wide column, or even several lines to describe *program content.* You will need detailed information here to secure a verification report from some stations. Also, allow ample room for a *remarks* column to jot down anything special about the report. Maybe it's the first time you've heard the station or received signals from that part of the world, or perhaps you've tried out a new piece of equipment and want to record its performance.

Finally, use two narrow columns (*sent* and *received*) under a main heading, *verification.* With check marks under the appropriate columns, you will know at a glance whether you sent or received a verification.

30

Composing a Signal Report

An exciting sideline of shortwave listening is receiving QSL cards and verifications. These colorful, exotic reports from all over the world are fascinating to show to friends and family. But to win them, you must send a carefully prepared report to the station you have monitored. More than a half-dozen items concerning the broadcast should be checked out before you mail your report.

Determine the *date* and *time* you heard the station. If the station is engaged in international broadcasting, then give the time in GMT. Be careful of the date, however, since it may be either the next day or the preceding day when you convert your time to GMT. For example, if your time is July 23, 9:00 p.m., Eastern Standard Time, it's five hours later in Greenwich, England, or 2:00 a.m. of the next day, July 24.

If you're reporting to a station which broadcasts locally, then state the station's local time.

Next, give the *frequency* on which you monitored the broadcast. Be as accurate as possible, using your receiver dial, the announced frequency, or the frequency given in a program schedule. Also give the *band* in meters.

Table 4-1. SINPO Code for International Signal Reports

S Signal Strength	I Interference	N Atmospheric Noise	P Propagation Disturbance (Fading)	O Overall Merit
5—Excellent	5—None	5—None	5—None	5—Excellent
4—Good	4—Slight	4—Slight	4—Slight	4—Good
3—Fair	3—Moderate	3—Moderate	3—Moderate	3—Fair
2—Poor	2—Severe	2—Severe	2—Severe	2—Poor
1—Barely Audible	1—Extreme	1—Extreme	1—Extreme	1—Unusable

Give as detailed a description of the *program content* as you can. Include such items as the name of the program, names of people heard, titles of musical selections, specific quotes from voices heard, and anything else to prove you actually received the station.

Give a rundown on your *receiving equipment,* type of antenna used to pick up the transmission, and a brief description of your receiver.

The item that will probably spark the most interest in your letter will be the *signal report.* Personnel at international stations are keenly interested in how well their signals are heard throughout the world. There are two popular codes for rendering precise reports: SINPO (Table 4-1), and RST (Table 4-2). Use SINPO for reporting to international broadcasters, and RST for reporting to amateur (ham) radio operators.

Consider how SINPO is used. As you can see, each letter represents five receiving conditions: *S*trength, *I*nterference, atmospheric *N*oise, *P*ropagation disturbance and *O*verall merit. Each signal condition is judged and assigned one of five different degrees of quality. For example, let's say you're listening to the BBC and signal strength is fair (3), but there's no inter-

Table 4-2. RST Code Used to Report to Amateur (Ham) Operators

Readability	Strength	Tone (Code Only)
1—Unreadable	1—Barely Perceptible	1—Very Poor Note
2—Barely Readable	2—Very Weak	2—Poor Note
3—Readable With Difficulty	3—Weak	3—Rough AC Note
4—Readable With no Difficulty	4—Fair	4—AC Note
5—Perfectly Readable	5—Fairly Good	5—Musical Note
	6—Good	6—Modulated Note
	7—Moderately Strong	7—Near DC Note
	8—Strong	8—Good DC Note
	9—Extremely Strong	9—Pure DC Note

ference (5) in the background. Atmospheric noise at the time is slight (4), but you're getting rapid fading which you consider severe (2). Evaluating all these factors, you might come up with an overall merit of fair (3). Thus your signal report would be "SINPO 35423."

The RST method of reporting is used by amateurs. It is similar to SINPO in that several signal qualities are judged and assigned a number, this time according to *R*eadability, *S*trength, and *T*one. You will rarely use "T" since a tone report is reserved for code (cw) signals. That is, unless you're practicing code and can copy a ham operating on cw.

The next item is supplying sufficient postage for the reply. This isn't required in every case, but it helps improve your rate of return. It's advisable to provide the postage by purchasing an International Reply Coupon. Your local post office can give further details.

If you don't hear from the station within three or four months, there are several possible explanations. Some stations will not answer the SWL, or you may have failed to include your name and complete mailing address (it happens). If the station is a rare one, try sending a complete reception report at least once more.

31

Write for Schedules

Many shortwave broadcasters supply detailed program schedules to worldwide listeners at no charge. Besides program times and frequencies, the literature is often filled with background information on the country as well as major events and personalities heard on the air. One attractive example is *London Calling* (shown in Fig. 4-1), a regular publication of the BBC. If you want to receive such literature, write to the following list of addresses and make your request. These broadcasters transmit in English and other languages and beam many of their programs to the North American continent:

BBC
P.O. Box 76
Bush House, Strand
London, W.C.2, England

Fig. 4-1. Typical shortwave program schedule.

Deutsche Welle
5 Koln 1
Hohenzollernring 62 POB 100444
Federal Republic of Germany

Radio Canada International
P.O. Box 6000
Montreal 101, Canada

Radio Prague
Prague 2, Czechoslovakia

HCJB (Voice of the Andes)
P.O. Box 691, Opa-Locka
Miami, Fla. 33054

Radio Australia
Melbourne 3000, Australia

Radio Sweden
Stockholm 1, Sweden

Swiss Broadcasting Corp.
Overseas Services, CH 3000
Berne 16, Switzerland

32

Broadcasts in English

A list of major international shortwave stations that transmit regularly in English to the North American continent (and elsewhere) is given in Table 4-3. This is just a small sample of the hundreds of broadcasts available each week on the international shortwave bands.

The exact frequencies and times given here, as in all shortwave transmissions, are subject to change. However, this list is very useful as an approximate guide to the many interesting programs that are available at various hours of the day and night. To convert GMT to local time, consult the information given next in Item 33.

Table 4-3. Major Shortwave Stations Broadcasting in English to North America

Time (GMT)	Station	Frequency (MHz)
0900	Radio Japan	9.505
1100	Haiti	9.770, 11.835
1200	BBC	15.260
1217	Radio Canada	9.625, 11.720
1245	HCJB, Ecuador	15.115, 17.880
1400	Radio Prague	15.240, 17.840
1430	Voice of Indonesia	6.105. 9.585
1445	Radio Mali	17.725
1530	Radio Nederland	17.810, 21.570
1620	Vatican Radio	11.705, 15.210
1930	Radio Kiev	6.020, 7.230, 9.710
2000	Radio Ghana	11.850
0000	Windward Islands	11.970
0100	Radio Habana	9.525
0100	RAI (Italy)	9.575, 11.810
0100	Voice of Spain	6.130, 9.760
0130	Radio Bucharest	5.980, 6.150, 6.190
0130	Swiss Broadcasting	6.120, 9.535, 11.715
0300	RAE (Argentina)	6.090, 9.690, 11.780
0400	Radio Sofia	9.700
0415	Radio Thailand	7.115, 11.910

33

Time Conversion by Country

Here is a convenient way to find the difference between GMT and local time anywhere in the world. First, find the general location in one of the six regions of the world as listed in Table 4-4. Then find the country, or one that is adjacent to it. The number after the country denotes the number of hours to add or subtract to GMT to convert to local time. For example, when it is 2200 hours GMT, it is midnight in Finland since this country is indicated with a "+2."

Table 4-4. Time Conversion Chart

Add or Subtract Numbers as Indicated to Convert GMT to Local Time			
Africa, Mediterranean, Middle East			
Algeria	0	Malawi	+2
Angola	+1	Mali	0
Cameroun	+1	Malta	+1
Canary Islands	0	Mauritania	0
Chad	+1	Mauritius	+4
Cyprus	+2	Morocco	0
Egypt	+2	Nigeria	+1
Ethiopia	+3	Rhodesia	+2
Gambia	0	Seychelles	+4
Ghana	0	Somalia	+3
Iran	+3½	S. Africa	+2
Iraq	+3	Sudan	+2
Israel	+2	Syria	+2
Jordan	+2	Tanzania	+3
Kenya	+3	Tunisia	+1
Lebanon	+2	Turkey	+2
Liberia	− ¾	Uganda	+3
Libya	+2	Yemen	+3
Madagascar	+3	Zambia	+2
Canada, USA, and Mexico			
Canada		Mexico—Cont'd	
Newfoundland	−3½	Pacific Zone (PST)	−8
Atlantic Zone (AT)	−4	USA	
Eastern Zone (EST)	−5	Eastern Zone (EST)	−5
Central Zone (CST)	−6	Central Zone (CST)	−6
Mountain Zone (MST)	−7	Mountain Zone (MST)	−7
Pacific Zone (PST)	−8	Pacific Zone (PST)	−8
Yukon Zone (YST)	−9	Yukon Zone (YST)	−9
Mexico		Alaska (AST)	−10
Central Zone (CST)	−6	Hawaii (HST)	−10
Mountain Zone (MST)	−7	Bering Zone (BST)	−11

East and Far East			
Brunei	+8	Malaysia, E.	+8
Burma	+6½	Malaysia, W.	+7½
Cambodia	+7	Nepal	+5⅔
Ceylon	+5½	Pakistan, E. (Bangladesh)	+6
China	+8	Pakistan, W.	+5
Hong Kong	+8	Philippine Islands	+8
India	+5½	Singapore	+7½
Indonesia (Djakarta)	+7	Taiwan	+8
Japan	+9	Thailand	+7
Korea	+9	Vietnam, N.	+7
Laos	+7	Vietnam, S.	+8

Europe			
Albania	+1	Italy	+1
Austria	+1	Luxembourg	+1
Belgium	+1	Netherlands	+1
Bulgaria	+2	Norway	+1
Czechoslovakia	+1	Poland	+1
Denmark	+1	Portugal	+1
Finland	+2	Romania	+2
France	+1	Spain	+1
Germany	+1	Sweden	+1
Gibraltar	+1	Switzerland	+1
Greece	+2	USSR (Moscow)	+3
Hungary	+1	Yugoslavia	+1

Pacific			
Australia		Gilbert Islands	+12
Victoria, New South Wales, Queensland	+10	Guam	+10
N. Territory, S. Australia	+9½	Mariana Islands	+10
W. Australia	+8	Marshall Islands	+12
Caroline Islands	+10	New Caledonia	+11
Fiji Islands	+12	New Guinea	+10
		New Hebrides	+11
		New Zealand	+12
		Solomon Islands	+11

West Indies and Latin America			
Argentina	−4	Guyana	−3¾
Bahamas	−5	Haiti	−5
Barbados	−4	Honduras (Rep.)	−6
Bermuda	−4	Jamaica	−5
Bolivia	−4	Leeward Islands	−4
Brazil, E.	−3	Nicaragua	−6
Brazil, W.	−4	Panama	−5
British Honduras	−6	Paraguay	−4
Chile	−4	Peru	−5
Colombia	−5	Puerto Rico	−4
Costa Rica	−6	Surinam	−3½
Cuba	−5	Trinidad	−4
Ecuador	−5	Uruguay	−2½
Falkland Islands	−4	Venezuela	−4
Guatemala	−6	Windward Islands	−4

Note that GMT is usually quoted by the 24-hour clock, where 10 p.m. is shown as 2200 hours, 11 p.m. is 2300 hours, etc. Also, this chart does not account for daylight saving time. If such time is in effect in a particular country or area, you must make an adjustment by first converting local time back to standard time. This is done by subtracting one hour from daylight saving time, then consulting the chart.

34

QSL Card Holder

QSL cards are a fascinating display to have on a wall, but consider these problems before you hang them. If you fasten cards to a wall, you obscure one side of the card, which often contains information that is just as interesting as that on the front. Also, you may damage a precious QSL card by tacking or taping it.

You can use your own ingenuity to get around the problem, but it's probably easier to obtain an item called a *QSL card holder.* This is a plastic sandwich that holds about 60 cards, and permits you to view both sides. It costs about a dollar and is generally available where ham or CB products are sold.

35

Receive More QSL Replies

Most of the big international broadcasters will send you a QSL card in response to a signal report at no charge. In some instances, however, the low operating budget of some stations may delay or even preclude the sending of a card unless you furnish return postage. A popular method to do this is with the *international reply coupon.* If you suspect that postage might hold up the card, you can pay for it in advance by using the coupon. They are available at many U.S. post offices. Do not send U.S. stamps since they may not serve for return postage.

If you have a local stamp dealer, you might also try this approach: purchase current stamps of the desired country, and send them along with your reception report.

36

Sked Timer

If there's a program you don't want to miss, and you worry about forgetting to turn on your receiver, try using an automatic timer. There are many inexpensive, electrically operated timers now available that can be used to turn on the receiver automatically at a preset hour. These timers plug into an ac wall outlet and have a socket that receives the plug from a lamp or any electrical appliance (see Fig. 4-2). You can usually preset such timers to turn on the power up to about 23½ hours in advance. When using a timer, leave the receiver volume control on normal listening level, or slightly higher, so that you will be sure to hear it and know that a program is about to be broadcast. If you have a tube-type receiver, adjust the timer so that it turns on the set about 20 minutes earlier than the scheduled program time. This allows the circuits to stabilize and have the least amount of tuning drift when the program comes on.

Fig. 4-2. An automatic timer.

Courtesy Radio Shack

37

Reception Guide

Here is a general reception guide (Table 4-5) listing the various bands and times for hearing different regions of the world.

Table 4-5. Reception Guide

Band (Meters)	Fre-quency (MHz)	Time	Zone
80	3.5	Morning, Evening	Local (Amateur)
49	6	Evening	Latin America and Europe
41	7	Evening	Europe
40	7	Late Afternoon, Evening	Europe
40	7	Morning	United States (Amateur)
31	9	Morning	Asia and Australia
31	9	Afternoon	Europe and Africa
31	9	Evening	Europe and Latin America
25	11	Morning	Asia and Australia
25	11	Evening	Latin America
20	14	Late Morning, Afternoon	United States, Foreign, (Amateur)
19	15	Late Morning, Afternoon	Europe and North America
19	15	Evening	North and Latin America
16	17	Afternoon	Europe
	17	All Day	United States
	17	Evening	South America
13	20	Afternoon	Europe
	20	All Day	United States
	20	Evening	South America
11	27	All Day	Local (Citizens Band)
10	28	Morning	Europe
	28	Evening	Central America, United States (Amateur), Asia

No chart of this type can be precise, because radio conditions change with the season and other variable factors. This one should prove most accurate during the spring and fall. During winter months, reception tends to be better on the lower frequency bands, while in the summer, reception is usually better on the higher frequencies.

SECTION 5

Techniques

38

Tuning Single Sideband

Because of crowding in the shortwave bands, many stations have switched from conventional a-m transmission to ssb, or single-sideband transmission. This system not only makes more effective use of transmitter power, but occupies about half the bandwidth of a conventional a-m station. Single sideband is widely used in the phone (voice) portion of ham bands such as 75, 40, and 20 meters. Single sideband is also assigned to the marine band between 2 and 3 MHz, and many military stations also use it. When tuning an ssb signal, you'll hear only "gibberish" unless special tuning techniques are used.

The problem is that an ssb station transmits no carrier signal. This must be replaced by the receiver in order to recover audio intelligence from the transmitted signal. For best results, the receiver should have an agc on-off switch, an rf gain control, and a bfo (beat-frequency oscillator). You can do without the first two controls, but the last item is a necessity since it reinserts the missing carrier. An "S" meter is helpful as a tuning guide.

When you hear the garbled audio of a single-sideband signal, turn off the agc, reduce the rf gain, and increase the audio volume. The best results are obtained when the volume is turned fully up, with the rf gain set just high enough to read the signal. Rock the fine tuning control, if you have one, to center the signal in the receiver. Determine the strongest reception by ear or "S" meter. Once the receiver is set up in this fashion, turn on the bfo. Slowly turn the bfo control through the signal and listen for a change in audio pitch between high and low tones, even while the words are still unintelligible. Concentrate on the lower tones, then very carefully search for a setting of the bfo that clears up the speech and makes it understandable. This is a very fine point in tuning, so keep trying if the correct bfo setting is hard to locate. Once you have the station properly tuned, you'll find that ssb signals are often free of static and interference. (A practical bfo that you can build is described in Section 3.)

39

Tuning Morse Code

Although the use of International Morse code has declined as a means of communications, it may never completely disappear from the shortwave bands. It is primitive compared to high-speed teletype (which has largely replaced it), but no other medium can so reliably transmit intelligence between two distant points with so little power and equipment.

The technical term for almost all code transmission is cw (for "continuous wave"), and receiving it requires a receiver equipped with a bfo (beat-frequency oscillator). Code reception, in fact, is similar to receiving single sideband, which was described in Item 38.

You might ask, "Why bother to receive code?" The best reason is that it can help prepare you as an avid SWL to become a ham operator. The shortwave receiver can be tuned to the frequent code-practice sessions broadcast on amateur bands by the American Radio Relay League. (Any issue of the magazine *QST* contains the operating schedule and frequencies.) Another reason for learning code is to identify unknown signals that have no voice modulation. These include beacon and other radionavigation stations. You will not need any special tuning techniques for many of these stations, however, since they often use mcw (for "modulated continuous wave"). Such stations are simply adjusted for maximum volume, as in ordinary a-m reception.

Yet another reason for learning code is to copy weather and other traffic still conducted on cw. This takes considerable experience, however, since commercial operators often send and receive at speeds from about 16 to 20 words per minute. It is excellent practice, though, if you're preparing to take the code test for a ham license.

To tune code on a receiver, turn off the agc and turn on the bfo. The bfo converts the otherwise inaudible code signals into audible tones. When the signal is strongest in the receiver, try to use the least rf gain and maximum audio volume. If the receiver has a system for increasing selectivity, adjust it to a narrow position to reject interference. Slight changes of the bfo control vary the pitch of the tone for the most comfortable listening. With some experience, you should be able to read one code station, while simultaneously receiving even two or three others.

40

Using a Selectivity Control

If your receiver has a control (switch) for adjusting selectivity, you've probably noticed its effect on audio fidelity. The reason for this is that the control changes the bandwidth of the i-f amplifiers in the receiver. When the bandwidth is narrowed, the higher audio frequencies are reduced in amplitude or even rejected entirely, depending on how much the bandwidth is narrowed. The degree of selectivity is usually chosen according to the type of signal being received and the existing band conditions. Most receivers usually have three positions (degrees of selectivity) on the selectivity switch (see Fig. 5-1). Although your receiver may not be calibrated precisely in the same manner, here are the major selectivity positions:

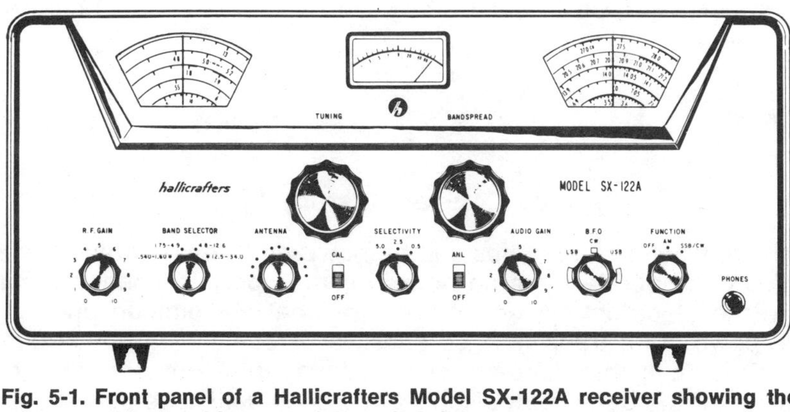

Fig. 5-1. Front panel of a Hallicrafters Model SX-122A receiver showing the three-position selectivity switch (bottom center of panel).

1. *5 kHz*—This is the "wide-open" position, most often used when fidelity is important. You would use this position when tuned to music or entertainment broadcasts.
2. *2.5 kHz*—This is an intermediate position and is the recommended setting for receiving single-sideband signals.
3. *0.5 kHz*—This is the narrowest selectivity position, and is the one best suited for code (cw) reception.

The advantage of increased selectivity is that as the receiver's bandpass is made sharper, the rejection of interference from other stations, as well as from natural and man-made sources, is increased. For this reason, you should consider listening to entertainment programs with some loss of fidelity, if necessary, by narrowing the selectivity from 5 to 2.5 kHz. The sharper setting may reduce interference to a program that is otherwise ruined by interference.

41

Learn a Foreign Language

A shortwave receiver affords a splendid opportunity for learning a foreign language. You hear native voices and accents directly from the country of origin. Not only is this an excellent supplement while learning a language in school, but many stations conduct complete language courses. Some of the major international broadcasters and the languages they teach are: Radio Japan (Japanese), Radio Portugal (Portuguese), Radio Sweden (Swedish), Deutsche Welle (German), Radio Nederland (Dutch), and Radio Australia (Indonesian, Thai, and Vietnamese). English lessons are given by the BBC and Voice of America.

42

Get Longer Battery Life

Many solid-state radios can be powered three different ways: You can operate them on house current, portable batteries, or on the electrical system of a car or boat. A common problem is powering a transistorized shortwave receiver on house current while its internal batteries (assuming they are not rechargeable) slowly deteriorate. Not only do such cells lose power over a period of inactivity, but they may also leak and damage the inside of the radio. An effective way to overcome both problems is to remove the batteries while using house current, and put them in cold storage.

A major manufacturer has conducted tests which reveal that the shelf life of batteries is greatly extended when the cells are stored at low temperature. The reason is that chemical reactions which consume the battery are slowed. A test made on standard carbon-zinc cells showed that cells stored at 9° C were in better condition at the end of *five* years, than those stored at 40° C after only *one* year.

The investigation also suggested practical techniques. The batteries can be stored at freezing temperatures if you avoid moisture condensation on the cells. Also, you must allow sufficient time for the batteries to warm up before placing them back in service. One method for preventing moisture is to wrap the batteries in plastic sheeting (household wrap will do) and place them in a home freezer. This reduces temperature to nearly 0° F, and no trouble has been experienced with cracked battery jackets. If no freezer is available, the cells can at least be placed in a refrigerator.

43

Adapter for Recording Programs

This easily assembled adapter enables you to record programs off the air without interrupting the sound in the speaker. You can leave the adapter connected to the tape recorder all the time if you wish, which makes it convenient when you hear something special (like a distress message) and want to make a recording quickly.

As can be seen in Fig. 5-2, the adapter is simply a "patch cord" made up of a length of shielded audio cable, a phono or phone plug, and a pair of alligator clips. You can do without the alligator clips if you would rather leave the adapter per-

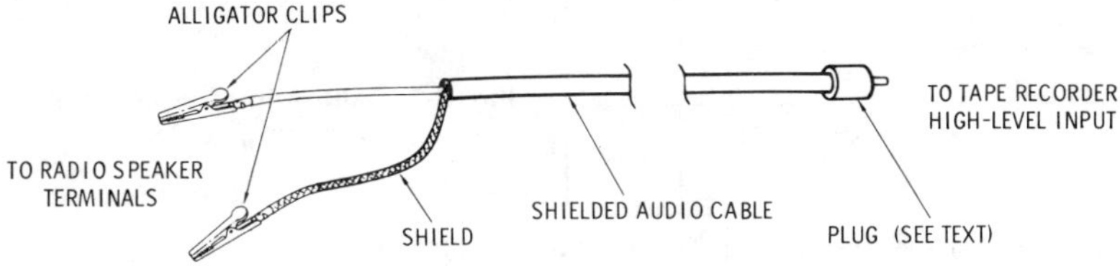

Fig. 5-2. Recording adapter.

manently connected to your receiver. The end of the cable marked "plug" is fitted with the type of phono or phone plug that mates with the input jack on your tape recorder. The other end of the cable is connected to the speaker terminals in your receiver. Either use the alligator clips for this, or solder the leads permanently to the speaker terminals. Before soldering the leads, however, you should test the hookup to find the correct ground connection for the cable shield at the speaker terminals. If you experience loud hum through the tape recorder, reverse the connections to the radio speaker.

Use a high-level input on your tape recorder for the plug end of the adapter. This is usually any input that is not for a microphone. Such inputs may be labeled TUNER, AUXILIARY, LINE, RADIO, TV, PHONO, etc.

To adjust for correct recording level, first tune in a station and adjust the volume control on your receiver for normal listening level. Then adjust the recording-level control on the tape recorder until the recording indicator shows satisfactory audio level.

44

Using a Logging Scale

On the bandspread dial of many shortwave receivers there is a scale marked 0-100 for logging purposes. The numbers do not represent frequency values, but allow you to adjust the bandspread capacitor to quickly find a specific frequency. Despite the approximate function of the logging scale, it is possible to turn it into a useful tuning reference. It enables you to come very close to a desired frequency in a shortwave band. While it will not take the place of the crystal calibrator described elsewhere, it is worth the effort to prepare your own calibration scale to convert log numbers into actual frequencies.

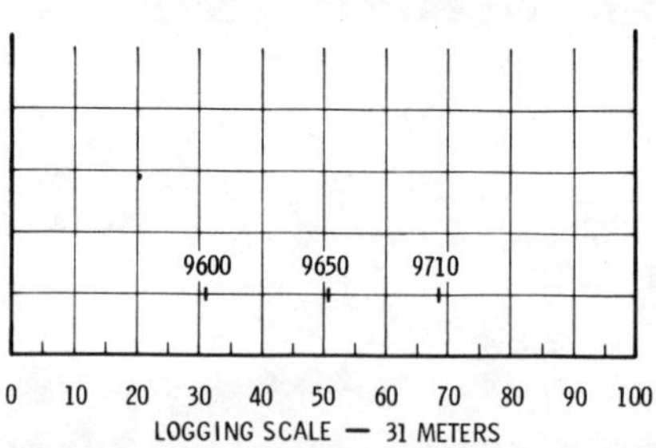

Fig. 5-3. Calibrating frequencies on a logging scale.

The scale is plotted on a piece of graph paper, or you can rule your own paper. Set the main tuning dial on the lowest frequency you can find in a given shortwave band. Place the bandspread on the zero point of the logging scale. By referring to shortwave program schedules, or listening to voice announcements over the air, you should be able to identify the frequencies of several incoming signals. Each time you can identify a frequency, write its value above the equivalent point on your graph, as shown by the three examples in Fig. 5-3. Similar scales can be prepared for other bands, but usually the higher bands need it most since tuning is increasingly critical at the upper end.

For your logging scale to be accurate at a later time, you must always reset the main tuning dial to the original point used for calibrating. A crystal calibrator is a helpful accessory to use with a logging scale since it can spot precise frequency points every 100 kHz on the graph.

45

Radiotelephone Talk

Broadcasters transmit in plain language, but stations in two-way communications resort to special phrases and phonetics. Band conditions are not always favorable and it's easy to hear a "d" as a "t" or confuse the numbers "five" and "nine." Certain words are standardized so their precise meanings are understood by all radiotelephone operators. The familiar "roger" at the end of a message, for example, is actually a phonetic for the letter "r," which means "received." It is understood that "roger" is never given unless all parts of the message, with no omissions or breaks, are received. If you wish to monitor the communications services which populate shortwave bands—boats, planes, the military, and others—consider some phraseology to make them intelligible.

Although the radio operator on a sinking ship in a tv drama may say: "Roger, wilco, over and out," it would hardly be heard anywhere else. A message usually terminates in only one of those words. The term "over" means the transmission (not message) is ended and a reply or return transmission is expected from the other station. If the message ends with "out," then no reply is expected. "Wilco" is short for 'will comply" and means just that. The words "yes" and "no" are usually replaced with "affirmative" and "negative."

Table 5-1. Phonetic Alphabet

Letter or Number	Phonetic Word	Pronunciation*
A	Alpha	*AL*-FAH
B	Bravo	*BRAH*-VOH
C	Charlie	*CHAR*-LEE (or *SHAR*-LEE)
D	Delta	*DELL*-TAH
E	Echo	*ECK*-OH
F	Foxtrot	*FOKS*-TROT
G	Golf	GOLF
H	Hotel	*HOH*-TEL
I	India	*IN*-DEE-AH
J	Juliett	*JEW*-LEE-*ETT*
K	Kilo	*KEY*-LOH
L	Lima	*LEE*-MAH
M	Mike	MIKE
N	November	NO-*VEM*-BER
O	Oscar	*OSS*-CAH
P	Papa	PAH-*PAH*
Q	Quebec	KEH-*BECK*
R	Romeo	*ROW*-ME-OH
S	Sierra	SEE-*AIR*-RAH
T	Tango	*TANG*-GO
U	Uniform	*YOU*-NEE-FORM (or *OO*-NEE-FORM)
V	Victor	*VIK*-TAH
W	Whiskey	*WISS*-KEY
X	X-ray	*ECKS*-RAY
Y	Yankee	*YANG*-KEY
Z	Zulu	*ZOO*-LOO
1	One	WUN
2	Two	TOO
3	Three	TREE
4	Four	*FO*-WER
5	Five	FIFE
6	Six	SIX
7	Seven	*SEV*-EN
8	Eight	AIT
9	Nine	*NI*-NER
0	Zero	*ZE*-RO

*Syllables to be emphasized are in italics

A phonetic alphabet devised by the ICAO (*I*nternational *C*ivil *A*viation *O*rganization) is widely used on the airwaves. It replaces older systems to make phonetics more universally understandable to persons who don't speak fluent English. (The complete phonetic system is shown in Table 5-1.) Several numbers are intentionally distorted in pronunciation to make them unmistakably clear over the air. The number "three" is given as "tree," "four" is lengthened to "fower," "nine" is given an extra ending and spoken as "niner," and "five" is shortened to "fife" to avoid prolonging the word and making it sound like "nine."

SECTION 6

Listening Away From Home

46

Traveling Antenna

Shortwave listening can be an added pleasure on a summer vacation (especially since a cabin or cottage may have no tv set). If you're camping out, a shortwave receiver might be the only source of electronic entertainment. A lot of listening can be done using a small whip antenna, but a good outdoor installation will pull in far more signals. Since a vacation antenna is only temporary, it can be built with few components and little effort. It doesn't have to withstand yearlong weathering or buffeting from the wind. The complete antenna travels easily since it's wound on a small spool.

The details are given in Fig. 6-1. The wire is No. 24 enamelled-copper type (also known as "magnet" wire) and is generally sold through electronic catalogs and supply houses in ½-lb spools. This weight provides a total wire length of 253 feet. Your antenna should run from about 70 to 100 feet but this depends on the available space and support points.

Two insulators are seen at the ends of the antenna. These can be cut from almost any plastic material. Drill two holes in each piece about an inch apart and attach the antenna wire in one hole and a length of strong cord in the other. The cord enables you to attach the top of the antenna to a convenient surface at the far end. At the near end, close to the receiver, you should have a small screw eye for a convenient tie point into a wood surface, if you need it.

The downlead of the antenna (a continuation of the top part) can be fed indoors through a piece of spaghetti tubing. The

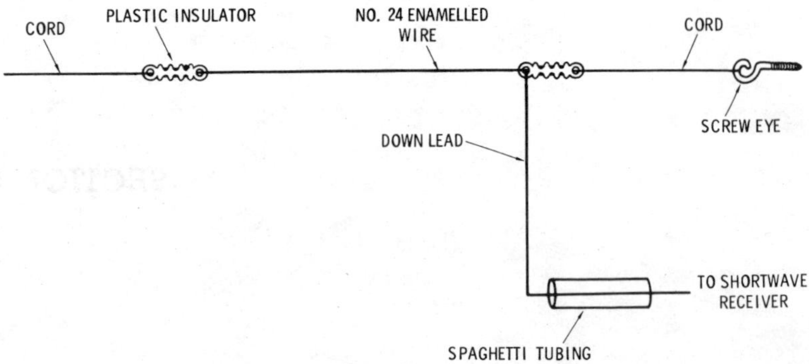

Fig. 6-1. A traveling antenna.

tubing prevents chafing the wire where it goes through a window. When the wire is connected to the shortwave receiver, first scrape away the enamel to the bare copper, but avoid any nicks that could cause a break. When vacation is over, just roll the antenna onto a spool and use it again next year.

47

Mobile Converters

A device used by amateur radio operators—the mobile converter—is also available to the shortwave listener. Instead of buying a complete receiver to cover special bands, you can obtain a converter which transforms the higher shortwave frequencies down to a standard broadcast-band frequency. The signals are then fed into the regular car radio and are heard through the regular car speaker. The converter also contains a changeover switch that enables the existing antenna on the car to serve both shortwave and standard broadcast reception. A converter like the one shown in Fig. 6-2 is available in models that cover the marine band from 2 to 2.85 MHz, and another model that covers international shortwave bands between 9.5 and 16 MHz.

Installing a converter is quite simple and requires no difficult wiring or mounting. The converter case is fastened to the underside of the dashboard at a position within easy reach, so that you can tune the dial while driving. The plug from the antenna is unplugged from the car radio and inserted into a mating socket on the converter. A new antenna plug from the converter is then inserted into the car radio.

Courtesy Tompkins Radio Products

Fig. 6-2. A mobile converter.

Power for a converter may be from its own internal battery. This simplifies installation and avoids interference often radiated through the power wiring of some automobiles. In other models, the converter taps into the 12-volt electrical system of the car. After converter installation, the car radio must have its antenna trimmer readjusted since some capacitance is introduced into the antenna line. This is quickly done by following the procedure detailed in Item 48, "Sharing a Car Antenna."

When a converter is switched on, the car antenna feeds the shortwave signals into the converter, and the converter signal is then fed into the car radio. You must tune the car radio to a frequency recommended by the manufacturer (usually around 1500 kHz). You may also set up one push button on the car radio to the desired converter frequency. This, however, removes one push button from service when you're listening to standard a-m broadcast stations. You can tune the radio manually to the converter frequency almost as easily as using a push button. This point is not difficult to locate since you will hear a distinct hiss in the speaker when you have tuned to the right frequency for converter reception.

48

Sharing a Car Antenna

The shielding provided by the metal body of a car prevents satisfactory reception of shortwave signals if you use the shortwave receiver's built-in loop or whip antenna. One solution is to mount a second whip on the car and connect it only to the shortwave set. A simpler method is to construct a small switch box that enables you to connect the existing a-m whip antenna to either the car radio or the shortwave receiver. (You probably won't be listening to both receivers at once, anyway.)

The details of such a device are shown in Fig. 6-3. It mostly contains a switch and a socket, so a small metal "minibox" of just a few square inches will do. You'll need one automobile antenna socket and plug, a single-pole, double-throw (spdt) switch, and a few feet of coaxial cable. The socket and plug are often listed in catalogs as the "Motorola" type; the switch may be the common slide type, or any small type for radio-frequency operation. Voltage and current capacity are not important here, so don't use a heavy switch intended for ac power.

Start by locating a convenient position under the car dash for the switch box. It should be within arm's reach of the driver and

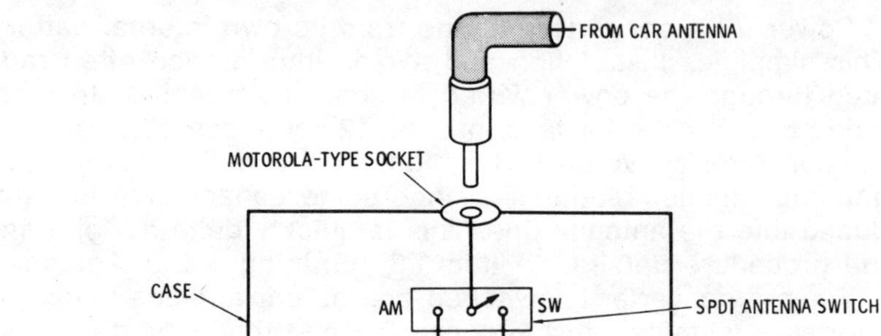

Fig. 6-3. Switching system for sharing car antenna.

clear of such items as the ashtray. A good position is to the side of the car radio that receives the cable from the car antenna. Feel behind the radio to determine where the antenna plug is and pull it out. It will be reconnected into the switch box later.

Before you do any wiring, mount the socket on the rear of the box and mount the switch on the front. Then bring the cable that will be connected to the car radio into the box and solder its shield to the metal case of the box with a solder lug and a nut and bolt. Connect the inner conductor of this cable to one outside pole of the switch, as shown in Fig. 6-3. Solder a Motorola-type plug on the other end of this cable after you have determined the length needed to connect between the switch box and the car radio. Connect the inner terminal of the Motorola-type socket to the center pole of the switch, as indicated in Fig. 6-3. The type of connection to the shortwave receiver depends on the antenna connections to the receiver. If there are two antenna terminals on the shortwave set, connect a shielded cable between them and the switch box, with the inner conductor going to the other outside pole of the switch as shown in Fig. 6-3. If there is only one terminal for the antenna input to the shortwave receiver, simply connect a length of hook-up wire between it and the switch.

Mount the completed switch box to the under lip of the dash with a nut and bolt. Plug the cable from the car antenna into the socket at the rear of the switch box, and plug the new cable from the "AM" side of the switch into the antenna socket on the car radio. The type and length of antenna lead to the shortwave receiver must be determined experimentally, as mentioned before. With the shortwave radio on, and the antenna switch set to the "SW" position, try both shielded and unshielded leads, with and without ground connections, to test for best results. Tune in several shortwave stations while you do this, and make sure you have the best type of antenna connections for the favorite bands you wish to listen to.

The final step is to retrim the a-m car radio. By inserting the switch box into the antenna lead, you have slightly changed the capacitance of the system. This is easily corrected by adjusting a variable capacitor on the car radio. This capacitor is located either next to the antenna socket on the car radio (look for a screw head in a large access hole), or at the front of the radio behind the tuning knob. In the latter case, the tuning knob must be removed to gain access to the trimmer capacitor.

On some car radios there will be instructions for adjusting the trimmer pasted on the radio case, but the procedure is almost always the same. First, tune the radio to a weak station on the high side of the a-m broadcast band (at about 1400 kHz). With a small screwdriver, carefully adjust the trimmer until you hear the maximum audio volume. Be sure the antenna switch is set to the "AM" position when making this adjustment.

49

Car Adapter for 9-Volt Radio

Many of the larger, solid-state portable shortwave receivers operate on a 12-volt battery and may be directly connected to an automobile battery. A special cable with a plug to match the car's cigarette lighter may be provided. It's also possible to operate the smaller, 9-volt portables in the car with a simple dropping resistor.

If you plan to listen to shortwave in a car on a 9-volt portable, the best power source is the radio's own battery. This avoids connecting into the automobile ignition system, which is often a source of considerable electrical interference. Noise voltages appear along the power lead to the radio and either radiate interference through the air or in the power wiring inside the radio. The degree of interference depends on the received frequency and the specific car. Generally, the lower the frequency, the more interference is heard. Operating the portable on its own batteries avoids this.

If, however, you must power the shortwave radio from a car battery, the voltage must be stepped down before applying it to the 9-volt radio. Although the ignition system of a car is called a "12-volt system, the actual voltage is closer to 14 volts while the car engine is running at a fast idle speed or higher. With a vom (volt-ohm-milliammeter) you should be able to quickly determine the value of dropping resistor required for your specific radio.

First, operate the radio at normal volume on its own battery. Then, measure the total current being drawn from the battery. This is done by connecting the probes of a milliammeter in series with one of the battery clips as shown in Fig. 6-4. Be sure to place the negative probe of the meter to the negative terminal of the battery or the reading will be reversed. During this check, the radio will continue to play and cause the meter to fluctuate. Observe the average current during a sustained passage of music and use this figure in your calculations. Let's say the figure is 20 mA.

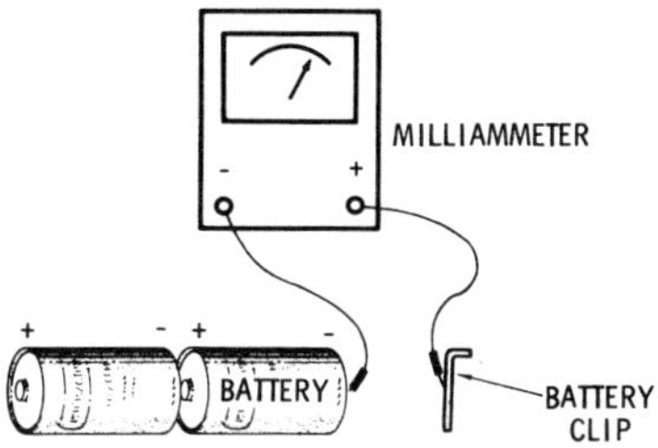

Fig. 6-4. Measuring total current drain from transistor-radio battery.

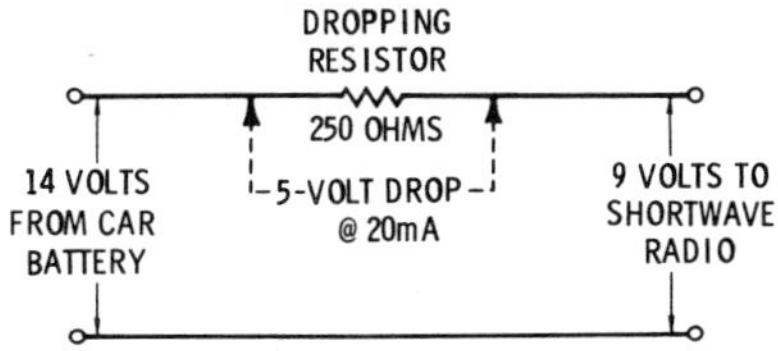

Fig. 6-5. Dropping resistor allows 9-volt radio to be operated from car battery.

Since the car battery measures about 14 volts, and the radio circuit requires only 9 volts, a total of 5 volts must be dropped across the resistor, as shown in Fig. 6-5. To determine the resistance (R) needed to drop this voltage, simply divide the voltage (E) by the current (I):

$$R = \frac{E}{I}$$
$$= \frac{5 \text{ volts}}{.02 \text{ amperes}}$$
$$= 250 \text{ ohms}$$

Note that the 20-mA current value must be converted to amperes by moving the decimal three places to the left. The answer, 250 ohms, is the resistance value.

One other calculation is required, and that is to determine the wattage rating of the resistor. This is found by multiplying the current by the voltage:

$$P = E \times I$$
$$= 5 \text{ volts} \times .02 \text{ amperes}$$
$$= 0.1 \text{ watt}$$

Thus, a standard ½-watt resistor should easily handle the slight amount of heat developed by the current through the resistor. If your radio draws considerably more current, always double the resistor wattage obtained by the formula to introduce a safety factor.

All that remains is a method for connecting into the electrical system of the car and feeding the voltage to the shortwave receiver. The most convenient (but temporary) method is through a plug which matches the cigarette lighter. Otherwise, a permanent connection can be made at the fuse block in the car by tapping voltages from almost any fuse. Connect to the side of the fuse that is away from the car battery and you will have the added advantage of fuse protection against a possible short circuit. Most portables draw little current and "stealing" power this way should not affect or overload the fuse.

Connect the 9-volt side of the adapter to the radio, being especially careful to observe battery polarity. This can be checked with the voltmeter. Also, check to make sure there are no reversed grounds. A car body is almost always negative, whereas the shortwave receiver might have a positive ground. In this case, a positive voltage would appear on the external metal parts of the radio. This can be checked by touching the voltmeter between the car body and the metal parts of the portable. If you read a voltage difference between these areas of 9 volts or more, then avoid touching the portable to the metal parts of the car. In this case, the radio must be kept in a plastic case or other insulator.

50

Restrictions Aboard Aircraft

Shortwave receivers are often so miniaturized that it is tempting to carry one aboard a commercial airliner. At 8 miles above the earth, reception might prove excellent. There are, however, definite restrictions on operating electronic devices aboard aircraft. There have been instances where a portable has disturbed the accuracy of a plane's navigational receiver or interfered with communications.

The source of the trouble is radio signals radiated by the local oscillator inside the shortwave receiver. Although by law such radiation must be limited by the manufacturer, the portable might be located within a few feet of the aircraft's antennas. The shortwave receiver does not have to be tuned to any of the aircraft's operating frequencies for such interference to occur. So do *not* play the set on *any* band while you're a passenger aboard an airliner. In the case of a private plane (Fig. 6-6), the same rule applies when the plane is flying "IFR" (under Instrument Flight Rules). The pilot can tell you this.

Not all electronic devices are excluded. The rule does not apply to portable voice recorders, hearing aids, heart pacemakers, and electric shavers. Other items might be permitted too, but this is decided by the operator of the aircraft.

Fig. 6-6. Navigational instruments in aircraft.

51

Aboard Ship

A portable shortwave receiver on the deck of an ocean-going vessel provides interesting listening. There are two limitations, though. One is that few signals are received inside the cabin of a steel-hulled ship. Some signal pickup can be obtained, however, if the whip antenna is held against the glass of a porthole or if a thin wire is taped to it.

The other limitation happens when the ship's radio is transmitting. Radio power of several hundred watts is common on large ships, and the signals may block your receiver. They cause false signals to appear at many points of the dial on several bands. The signals are often unintelligible since they're in Morse code, radioteletype, or single sideband.

If you're aboard an ocean liner, be sure to visit the radio "shack." On large ships, the shack is a floating shortwave station with several transmitters, receivers, switching panels, and other equipment. There should also be a small battery-operated rig for emergencies or power failure. A tour through this facility will be of considerable interest to any SWL. The ship's radio officer is often proud to show off his shack to visitors.

52

Direction Finding

A portable shortwave receiver with direction-finding capability is of practical value to the listener who enjoys boating and other outdoor activities. There are commercial receivers especially made for this purpose (one example is shown in Fig. 6-7). This model covers most of the major shortwave bands from 1.6 to 22.4 MHz, as well as the standard a-m and fm broadcast bands, a pretuned vhf/fm weather band on 162.55 MHz, and the longwave weather/navigation band from 150 to 400 kHz. The direction-finding capability is made possible by a special swivel base and a compass rose. The antenna used for direc-

Courtesy Zenith Radio Corporation

Fig. 6-7. A portable shortwave receiver that has special direction-finding capabilities.

tion finding is a standard loopstick mounted inside the receiver cabinet. It is only used on the standard a-m broadcast band and the longwave weather/navigation band. By turning the receiver on its swivel base and listening for a drop in station volume (null point), you can obtain a magnetic bearing to the station. The receiver also has a tuning meter which helps to locate the null point more accurately. These direction-finding features come in handy when cruising at night, in fog, or during other periods of limited visibility. You can either "home-in" on a radio station, or use a chart to obtain a "fix." Two lines of position are drawn on the chart, and where the lines intersect "fixes" your position. These techniques are covered in detail in the instruction manuals that accompany this type of receiver.

It is not generally known, but any portable transistor radio that uses a loopstick antenna for the a-m broadcast band can be used for direction finding in an emergency. You can quickly check this by placing the radio on a flat surface and rotating it while listening to a station. There will be two points during one complete rotation that cause the volume to drop severely. These are the null points, and are used in direction finding because they are much sharper and more easily pinpointed than are the two points of maximum signal reception. Try this first on an a-m broadcast station where you know the exact location of the transmitter tower. You will notice that as you reach the null points, the loopstick inside the radio will be oriented in such a manner that an imaginary line runs through the length of the loopstick's ferrite rod and through the transmitter tower of the station. Thus the loopstick becomes the pointer to or from the station. As in any direction-finding receiver, there is a double indication (an ambiguity) and you must know the correct *general* direction to the station. If you have a map and a compass, though, you will usually know this, even if you're lost.

The best stations for direction finding are in the longwave band, which covers from 150 to 400 kHz. These stations are shown on nautical and aeronautical charts. The standard a-m broadcast band is usually the only band on shortwave portables that operates from a built-in loopstick antenna. The other bands use a telescoping whip or other antenna, and cannot be used for direction finding.

53

Spark-Plug Suppression

The fun of listening to shortwave while in a car or boat is diminished by spark-plug noise created while the engine is running. This noise is easy to identify because it disappears when the ignition key is turned off. Also, the rate of the popping noise is in step with the speed of the engine. The vehicle may already have certain factory-installed items to reduce spark-plug noise, so identify them before taking any added measures. A popular method is to install special resistance cable between the plugs and the distributor. The cables may be marked with words to that effect. If your engine wires are not, you might be able to obtain a set and install them. The job isn't difficult in most cases, but be sure the cables are specifically made for your car. The resistance element inside the cable is cut to an optimum length.

Another technique for reducing spark-plug noise is to use special resistor-type plugs. They have built-in suppressor resistors and can replace your existing plugs. Yet another possibility is to install external suppressor resistors which are made to clip to the top of a conventional plug. In any installation, be sure all cables are free of cracked insulation since electrical leakage is also a source of radio noise.

A boat with a nonmetallic hull may prove very noisy to reception, and the steps mentioned above may not quiet it. One answer is the shielded ignition system. This is a kit with various cables encased in a flexible metal jacket, and metal containers for other components. This is considerably more expensive than the usual suppression measures, but if cost isn't a factor, it's an effective treatment. Special shielding kits are available for both boat and car engines.

54

Alternator Bypassing

After spark plugs, a vehicle's alternator is the leading cause of noise in mobile shortwave reception. Several measures prove helpful and these can be applied to a car or boat engine that is disturbing shortwave listening. You can obtain devices called

Fig. 6-8. Alternator bypass capacitor.

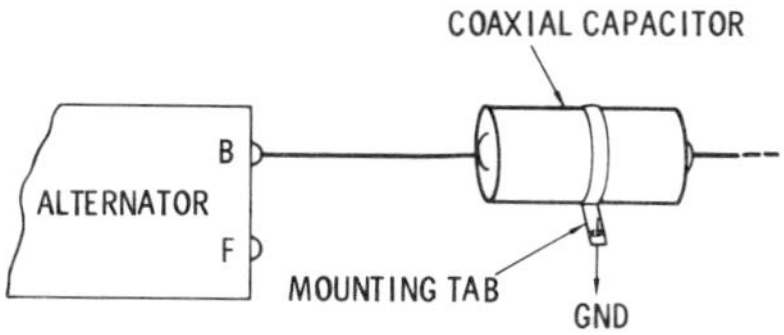

"alternator filters," which are usually bypass capacitors that ground the interference. You may also purchase high-quality *coaxial* capacitors with a rating of 0.5 μF (at 40 amperes) and achieve a similar effect. Unlike common paper tubular capacitors of the same rating, these are metal enclosed and furnish a better grounding path for noise voltages at higher frequencies. They are also self-mounting and are protected from the environment around an operating engine.

There are two points where the capacitor may be mounted; at the alternator terminal marked "B" (battery), or "F" (field) (Fig. 6-8). Try both positions to determine which affords the best noise reduction, or use a capacitor on each point, if necessary. The capacitor has a mounting tab which bolts to ground (on the alternator case, if convenient). Attempt to keep the capacitor as close as possible to the alternator for best electrical effect. The alternator wire is cut, and the free ends are connected to the capacitor end terminals.

55

Quiet Alternator

If you plan to do extensive shortwave listening in a car or other vehicle equipped with an alternator, consider these maintenance tips. They can reduce the degree of electrical interference from this source and have the side benefit of improving the alternator's primary function—to power the car's ignition system.

When the car is given an engine tune-up (about once a year in average driving), the alternator should be opened for inspection. There are slip rings inside which should be cleaned. The length of the brushes, and brush-spring tension, should also be checked. If the brushes appear worn, they should be replaced. If slip rings measure out-of-round, by even a small amount, this can raise the radio noise level and, just as important, impair the normal alternator operation.

SECTION 7

Special Signals and Bands

56

Identifying Strange Signals

Sweep through every band on a shortwave receiver and you will almost certainly hear strange sounds that have neither voice nor other quick means of identification. These signals are often powerful and, if tuned in frequently, are almost certain to arouse curiosity. Although it is difficult to describe some of these sounds in words, we'll try to identify a few of the most common ones.

Consolan—You'll pick up this one if you tune across the low-frequency band between about 150 and 500 kHz. Over a period of several minutes there will be a series of dots and dashes that slowly grow in volume, run together, then fall back into the background noise level. This is repeated continuously, taking just over one minute to complete each cycle. If you don't switch on a bfo, the sounds are audible only as a series of hisses.

These are the signals of the *consolan* system used by mariners to obtain accurate compass bearings over hundreds of miles. A special feature is that consolan can be used with any simple radio. The navigator merely counts the number of dots and dashes he hears, then consults a book of consolan tables to convert the number into a line of position to the station. One such station is on Nantucket Island, Mass., another is at San Francisco, with other stations located elsewhere in the world.

Loran—If you tune just below 2 MHz (or the band on most shortwave receivers that lies just above standard a-m broadcast), you will probably hear rapidly repeating pulses with the raspy nature of static. But they are so constant and reliable that they are obviously man-made. These are the signals of *loran* (for *LO*ng-*R*ange *A*id to *N*avigation) used by ships and aircraft navigating the high seas. The shortwave receiver picks up loran as a train of meaningless pulses, but the loran receiver displays them on an oscilloscope. The pulses originate from stations hundreds of miles apart, and the time difference between receipt of the pulses is used to locate lines of position on a loran chart, thereby establishing a navigational "fix."

Radioteletype—Much message traffic and press information travels via radioteletype, and the shortwave bands are loaded with such signals. When the receiver bfo is turned on, radioteletype sounds like high-speed Morse code. There are subtle differences in radioteletype signals which can often be detected by ear. By listening closely, you'll notice that unlike regular code, the radioteletype signal really consists of an audio tone which rapidly shifts up and down in frequency. Older systems use a shift of 850 Hz, while newer ones employ 170 Hz. It is impossible to read the radioteletype code by ear, but occasionally you'll detect patterns that seem to be repeating. This means that the station is probably sending out a general call (CQ) or transmitting a repeating stream of "Rs" and "Ys." These letters activate most of the teleprinter's mechanical parts, and are a good test of its operation.

Facsimile—This mysterious signal, which appears in many parts of the shortwave spectrum, is carrying "pictures by radio." Facsimile converts photographic images into a varying audio tone for long-distance radio transmission. You can often spot a photo transmitted by facsimile in a daily newspaper, because it will usually by identified by the words "radio photo" or "wireless photo" printed below it. The system operates by rotating the original picture on a drum and scanning it line-by-line with a photocell (as in constructing a tv image). This produces an audio tone whose pitch corresponds to the light and dark areas of the picture. When this signal is transmitted on the air, you can easily identify it since it sounds somewhat like a key being duplicated on the grinder in the local hardware store.

57

Hearing Distress and Emergency Calls

"Come at once, we've struck a berg. It's CQD, OM, (Come Quick, Danger, Old Man) . . . SOS . . . SOS . . . SOS . . ." That was the dramatic distress call sent out in Morse code by the radio operator of the ill-fated *Titanic* on an April evening in 1912. It wasn't the first emergency call ever transmitted by radio, but it was the first time the letters "SOS" were sent from a ship in distress. Even today an emergency call might be sent in similar fashion from a ship at sea. It would be heard on the international distress and calling frequency of 500 kHz.

Table 7-1. Major Distress Frequencies

Frequency	Description
500 kHz	International distress and calling frequency for radiotelegraphy. This can be picked up on receivers that have a low-frequency band just below the standard a-m broadcast band. Traffic is in Morse code.
512 kHz	Calling frequency for use when 500 kHz is occupied with distress messages.
2182 kHz	Marine international radiotelephone distress and general calling frequency. This can be found on shortwave receivers that have the medium-frequency band just above the standard a-m broadcast band. If the voices sound garbled, consider the signals to be single sideband (ssb), and tune them according to the instructions given in Section 5, "Techniques."
156.8 MHz	International distress, safety, and calling frequency for the maritime mobile radiotelephone service in the 156-174 MHz vhf band. Receivers that include this frequency usually have a band marked "PSB" (Public Safety Band), which includes frequencies from 147 to 176 MHz. The difference between this band and the medium-frequency (2 MHz) band is the type of activity. Pleasure boats are increasingly being assigned to the vhf band, while commercial mariners are concentrated on the medium-frequency band.
121.5 MHz 123.1 MHz	The frequency 121.5 MHz is a worldwide distress and emergency channel for aircraft, while the frequency 123.1 MHz is used by aeronautical search and rescue stations (air-to-air, and air-to-ground). These frequencies can be found on shortwave receivers having a vhf band covering from 108 to 136 MHz. Although distance is line-of-sight, a high-flying jet aircraft has a communicating range of about 200 miles.
27.065 MHz	As Channel 9 of the Class D Citizens band, 27.065 MHz is reserved for distress and emergency traffic.

With the growth of radio, many additional distress frequencies have been assigned throughout the spectrum. Most of the new ones use voice, not code, and can sometimes be heard on home shortwave receivers. Instead of the letters "SOS," an operator would say "mayday," which is the pronunciation of the French word "m'aidez" ("help me"). This is followed by a description of the emergency, the position of the craft, and other pertinent information such as the type of assistance required.

Emergencies are not frequently heard on the air, but some listeners occasionally monitor the distress channels. When an emergency breaks, it can provide very exciting listening. This writer has heard emergency calls on the aviation distress channel (121.5 MHz) several times. One evening during bad weather, for example, came the report of a pilot with a rough-running engine. His communications with a control tower grew more suspenseful by the minute, until he finally made his way to the field. Several other instances concerned lost pilots who were talked back home by air-traffic controllers on the ground.

One tip for anyone wanting to monitor an emergency: If you're watching television or listening to a regular radio, and a news story breaks about a ship in distress, rush to your shortwave receiver. Tune in one of the distress frequencies and you may hear live communications with the rescuers.

Table 7-1 is a list of major distress frequencies in use throughout the world.

58

Frequency Bands at a Glance

Not all shortwave receivers are calibrated with all the various broadcasting bands. If you wish to add such markings, Tables 7-2, 7-3, and 7-4 list the major bands of interest for shortwave listeners.

59

Decoding Ham Talk

The first time you tune in a ham radio operator, you'll find his conversation nearly unintelligible. Ham language is sprinkled with slang, abbreviations, and other inside talk. Actually, the

Table 7-2. Major International Shortwave Bands

Band (Meters)	Frequency (MHz)
120	2.300- 2.498
90	3.200- 3.400
75	3.900- 4.000
60	4.750- 5.060
49	5.950- 6.200
41	7.100- 7.300
31	9.500- 9.775
25	11.700-11.975
19	15.100-15.450
16	17.700-17.900
13	21.450-21.750
11	25.600-26.100

Table 7-3. Major Amateur (Ham) Bands

Band (Meters)	Frequency (MHz)
160	1.800- 2.000
80	3.500- 4.000
40	7.000- 7.300
20	14.000- 14.350
15	21.000- 21.450
10	28.000- 29.700
6	50.000- 54.000
2	144.000-148.000

Table 7-4. General-Interest Bands

Band	Frequency (MHz)
Citizens, Class D (hf)	26.965- 27.255
Citizens, Class A (uhf)	462.550-462.725
	467.550-467.725
Aviation (vhf)	108.000-135.950
Public Safety (mf)	1.610- 2.812
Public Safety (Low-Band vhf)	30.000- 50.000
Public Safety (High-Band vhf)	148.000-174.000
Public Safety (uhf)	451.025-467.525
Navigation and Weather (lf and mf)	.180- .420
National Weather Service (vhf)	162.550

number of baffling references is not very great and if you can decode a dozen or so, you'll probably understand much of the conversation. Consider this mythical transmission, and what the cryptic words mean (in parentheses):

CQ, CQ, CQ (general call to anyone who wishes to establish contact) from K1ODH (the ham's call sign—"K" refers to a U. S.-licensed station, while the numeral "1" means the New

England area). I read you Q5 (perfectly understandable), but there is some QSB (fading) on your signal. (Although these "Q" signals are really intended to shorten code transmissions, they're widely used in voice work, as well.) Have you heard any DX (distant stations) on 15 (meter band) lately? I worked a G (British station) last week but the QSO (contact or communication) was cut short by QRN (static, or interference from natural sources). By the time we finished the QSO, the QRM (man-made interference) really piled up. Maybe because it was a YL (young lady) giving 88s (love and kisses) to everyone on the frequency. Anyhow, my XYL (ex-young lady, meaning "wife") was calling me so I pulled the switches (turned off the station). Thanks for the QSO and 73 (best wishes).

60

Band and Frequency Converter

Here is a memory aid that quickly reveals how to convert from band to frequency and back. Shortwave stations may be listed in either meters (band) or megahertz (frequency), so it's of practical value to know how to change one into the other. Learn this simple device and you won't have to memorize three different formulas.

As shown in Fig 7-1, a circle is drawn with sections labeled 300, meters, and MHz. Place a finger over the value you wish to find, and the rest of the circle tells you what to do. For example, let's say you want to know the frequency of the 25-meter band. Since you wish to know frequency, cover MHz in the circle with a finger. Note that it leaves 300 *over* meters. Thus, it's telling you to *divide* 300 by 25. This gives an answer of 12 MHz, which is the approximate frequency of the 25-meter band.

If you wish to find meters, and you know the frequency, place a finger over meters. Now the circle indicates that you should divide 300 by the frequency in MHz for the answer.

Another variation is if you want to prove whether figures given for band and frequency agree with each other. Since you know

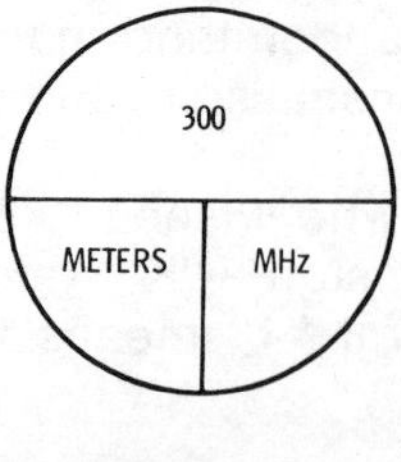

Fig. 7-1. Band and frequency converter.

these numbers, place a finger over 300. This places meters next to MHz in the circle in the standard mathematical arrangement to indicate multiplication. Thus, if the two numbers are correct, multiplying them should give you an answer of 300.

The figure "300" is a constant—the velocity of radio waves. It is actually 300,000,000 meters per second, but is shortened here to 300 since the frequency has also been shortened (1 MHz is actually 1,000,000 Hz).

61

Clandestine Stations

Seasoned shortwave listeners are very interested in a breed of broadcaster known as the "clandestine" (or secret) station. These stations almost always appear on the air during political upheavals, especially military coups and civil wars. They are illegal in that they do not observe international treaty agreements which regulate the airwaves. Such stations, however, are often exciting to hear since they represent history in the making. When newspapers or other media announce tension in some part of the world, consider it a possible prelude to choice shortwave listening. You'll have to search for it since station frequencies and schedules are rarely published.

Many clandestine stations, as the name implies, operate from secret transmitter sites. Since mobility can prove essential, a station may be portable and move among various locations to avoid detection. (Their signals can be pinpointed by direction finders.) Sometimes a clandestine station will operate outside the borders of the country it wants to influence. The content of such broadcasting is often political propaganda from revolutionary or exiled groups.

62

Spy Stations

If you tune across the shortwave bands, your curiosity may be aroused by an announcer reading numbers. The digits may occasionally be broken by meaningless words or phrases. You have probably stumbled across a "numbers" station, whose transmissions are the subject of lengthy discussion and controversy among long-time listeners. You will not be able to decipher the meaning of the numbers since they are secret messages to espionage agents located anywhere in the world.

The difficulty of cracking these codes has been demonstrated after the capture of spies who have used this system. It's done through a code book which translates the numbers back into letters. What accounts for the system's great secrecy is that each page of the code book is used only once, then destroyed. (The agents, incidentally, never answer their "numbers" stations since they could be detected by direction finders.)

63

Radio Pirates

The "pirate" is a broadcaster that the shortwave listener might encounter between the legal stations. Tuning and identifying the outlaw is often the delight of the ardent shortwave hobbyist. The pirate is a brigand who sails his ship just beyond the 12-mile territorial limit of a country to do his dirty work. The ship is a complete radio station including transmitter, antenna, studio, and talent. It broadcasts illicit programs to nearby population centers, which willingly receive them. The reason is that pirates send out very popular programming (usually music) that's not provided in sufficient quantity by the country's own stations. Unlike the U.S., most broadcast stations throughout the world are government-run and may load their schedules heavily with culture and propaganda programming. The pirate fills the vacuum and grows wealthy on commercial announcements bought by advertisers.

Despite their color and popularity, the pirates are often decommissioned by force and their advertisers pressured by law into withdrawing support. Nevertheless, radio piracy is a financially attractive profession and will probably continue to occur along the various coastlines of the world.

64

Bootleggers

The "bootlegger" is another interesting character the shortwave listener should be aware of. His stock in trade is to transmit radio signals over the air without benefit of a legal call sign, or even a license. One infamous bootlegger in the Baltimore area, for example, raised tremendous speculation among SWLs by calling himself the "Voice of the Blue Eagle." His broadcasts were variously called everything from a plot by a foreign power to the messages of a flying saucer. The station was ultimately revealed as a complete sham.

Bootleggers you're apt to hear will be on the standard broadcast band and conduct operations in your hometown. The Federal Communications Commission has a profile on the typical case since the agency receives more than two complaints a week from citizens about mysterious programs. An investigation usually exposes not a hardened criminal, but a youngster playing with a "wireless mike" or "phono oscillator." These are perfectly legal devices as long as they are operated according to the manufacturer's instructions. The trouble usually begins when an antenna longer than five feet is attached to the circuit, which greatly increases the transmitting range. Also, precocious experimenters add amplifiers to further boost the range. These miniature broadcast stations can reach all over the neighborhood, and can also be picked up on car radios.

Although the law provides stiff penalties for illegal broadcasting, the investigating engineer usually acts with a measure of understanding. An explanation to the youngster's parents about the seriousness of the operation is usually enough to restore peace to the airwaves.

65

Space Recovery

The avid shortwave listener can often witness the dramatic events which accompany the re-entry and recovery of space vehicles. Recovery forces at sea—ships and aircraft—often communicate on the high-frequency band that falls within the coverage of most shortwave receivers. The actual frequencies may lie approximately between 5 and 25 MHz. There is no exact guide to either bands or frequencies except that search and rescue operations may be heard on 20.007 MHz. A listener with a nimble hand, however, should be able to search these likely bands and check each voice transmission. With a bit of luck and patience, you'll discover the signals you want. It helps to know the exact time of re-entry and recovery by checking the newspaper or listening on regular radio or tv.

66

International Weather

There is a special group of frequencies in the high-frequency band (3-30 MHz) that carry aviation weather reports. They may prove valuable to SWLs who cannot tune the low-frequency weather broadcasts on the band (180-420 kHz) just below the standard broadcast band, or continuous weather information on the National Weather Service frequency of 162.55 MHz (vhf). These hf stations can be picked up on almost any shortwave receiver. The broadcasts are simultaneous on several widely separated frequencies, so you can select the one that produces best reception.

These hf weather broadcasts are prepared for pilots of transoceanic flights. They announce ceiling, visibility, temperature, dew point, and winds at major airports along the coast. Forecasts are also given. Although many of these details may not be of direct interest to the SWL, you may want to tune them during periods of severe weather (for example, when hurricanes or tornadoes are approaching the area). The broadcasts originate from Flight Service Stations operated by the FAA.

The following frequencies are in use for transatlantic flights and are transmitted from Kennedy International Airport in New York City. Since the frequencies are subject to long-distance skip transmission through the ionosphere, you should have little trouble hearing them at distances of several hundred miles, or even thousands of miles away. The frequencies tend to remain the same for long periods of time, but they are occasionally subject to change. (You should be able to locate these signals on the West Coast by scanning your receiver between 3 and 18 MHz.)

The Atlantic frequencies (in MHz) are; 3.001, 5.562, 8.868, and 13.272. The broadcasts begin every hour on the hour, and run 20 minutes past the hour. As New York stops transmitting, a Canadian station in Gander, Newfoundland, takes over (on the identical frequencies) and continues the reports from 20 to 30 minutes past the hour. Transmission then returns to New York for from 30 to 50 minutes past the hour. This sequence is then repeated. Since the New York and Gander transmitters are about 1000 miles apart, expect some change in signal strength as the stations switch. Choose an alternate frequency if you lose the signal.

67

Ship Communications

Ships plying international and other waterways are allocated a number of frequencies for long-range communications. As in the case of shortwave broadcasting, operators choose frequencies according to the distance required and ionospheric conditions. Thus, the assigned ship frequencies are well distributed throughout the high-frequency (hf) band. These channels are divided into "calling" and "working" categories. A calling channel is a meeting point where ship operators initiate con-

Table 7-5. Maritime Frequency Allocations (HF)

Type of Station	Frequency Band (MHz)
Ship (Telephony)	4.063 - 4.1395 6.200 - 6.2104 8.195 - 8.2812 12.330 -12.421 16.460 -16.565 22.000 -22.0945
Coast (Telephony)	4.361 - 4.438 6.514 - 6.525 8.7285 - 8.815 13.1075 -13.200 17.255 -17.360 22.6245 -22.720
Ship (Calling, Telegraphy)	4.178 - 4.187 6.267 - 6.2805 8.356 - 8.374 12.534 -12.561 16.712 -16.748 22.2225 -22.2675
Ship (Low Traffic, Telegraphy)	4.187 - 4.231 6.2805 - 6.3455 8.374 - 8.4595 12.561 -12.689 16.748 -16.9175 22.2675 -22.374
Ship (High Traffic, Telegraphy)	4.17225- 4.178 6.25825- 6.267 8.34175- 8.356 12.50325-12.534 16.6605 -16.712 22.1845 -22.2225

tacts with other stations. After communications are established, they move to a "working" channel. The system allows large numbers of stations to monitor few working frequencies and suffer the least interference from others engaged in message handling.

If you wish to monitor these channels, Table 7-5 lists some of the major frequencies set aside for the maritime services (ship stations as well as some coastal stations), together with the type of transmission. In general, frequencies are assigned to stations on the basis of type and purpose of station, geographic location, operating-range requirements, interference considerations, etc. In these bands, radiotelegraphy (code) is by on-off carrier keying, while radiotelephony (voice) is by single sideband (ssb). (Some conventional a-m may be encountered until the transition to ssb is completed.)

68

International Aircraft

Frequencies used by aircraft on overseas and intercontinental flights are allocated throughout the high-frequency (hf) band. Unlike vhf, which is widely used for air-traffic control within a country, these lower frequencies provide over-the-horizon transmission required for communications at great range. If you want to tune these channels, Table 7-6 lists the major aeronautical frequency bands. These channels are assigned throughout the world, and one frequency may be shared among several

Table 7-6. Aeronautical Frequency Allocations (HF)

Type of Service	Frequency Band (MHz)
Aeronautical Mobile	2.850- 3.155 3.400- 3.500 4.650- 4.750 5.450- 5.730 6.525- 6.765 8.815- 9.040 10.005-10.100 11.175-11.400 13.200-13.360 15.010-15.100 17.900-18.030
Aeronautical Fixed Aeronautical Mobile	21.850-22.000 23.200-23.350

regions. The frequency 8871 kHz, for example, is used by aircraft flying in Europe, North and South America, the Far East, the Caribbean, and Canada. Since communication distances are great, you may hear signals from areas well beyond your own.

69

Very Low Frequencies

The most intensive developments in radio have been in the higher end of the radio spectrum. Frequencies grow higher in the microwave region and communications satellites carry active microwave repeaters in orbit. But there is also considerable activity in the lower end of the radio spectrum. If a shortwave listener wishes to investigate this frontier of radio, he will encounter vlf (Very Low Frequency). Stations operating on vlf have no entertainment programs and, in fact, most transmissions are in code. Nevertheless, vlf attracts interest because it's an interesting medium and you can receive QSL cards from the transmitting stations by sending reception reports. The receiving equipment is simple.

The vlf band runs from about 10 kHz to 30 kHz, which means the signals are so low in frequency that they could almost be heard by the human ear (which has an upper limit of about 15 kHz). A vlf station, however, emits an electromagnetic wave that is inaudible and must be detected by a conventional radio receiver. The notable characteristic of vlf radio waves is that they saturate the earth's surface and mainly form a *ground* wave. There is very little "skip," as on the higher frequencies, and the waves are little affected by sunspots or other disturbances. The result is a communications medium that is extremely reliable on a daily basis.

A disadvantage of vlf transmission is its severely restricted bandwidth. The amount of information that can be modulated on vlf is limited since speech modulation would excessively broaden the bandwidth and reduce the number of stations that could occupy the band. This explains why most modulation on vlf is in the form of code (cw). Since code is merely an on-off operation of a radio carrier, it has the least bandwidth of all types of modulation. However, even code must be held to low rates of speed or the signal will occupy excessive bandwidth. This is an advantage to the listener who doesn't understand

Morse code; he can copy the dots and dashes as they are heard, then decode them at a later time.

Stations on vlf include powerful U.S. Naval installations in the million-watt category, a National Bureau of Standards station, WWVL on 20 kHz, located near Ft. Collins, Colorado, and other stations in the vlf band between 10 and 30 kHz. To receive these frequencies requires a special receiver. Since there is little available receiving equipment for these frequencies, hobbyists usually construct simple circuits from construction articles that appear in electronics hobby magazines. Vlf equipment is also available on the surplus market at moderate prices.

70

Geophysical Alerts

Since 1958, a world-wide system has been in effect to alert experimenters, researchers, and anyone else interested in radio propagation. It warns of such events as magnetic storms, solar flares, cosmic rays, and related natural phenomena that affect communications. Geoalerts are broadcast in voice during the 19th minute of each hour from WWV (Colorado), and during the 46th minute of each hour from WWVH (Hawaii). The transmissions are heard on one or more of the standard WWV frequencies: 2.5, 5, 10, 15, 20, and 25 MHz. Station WWVH broadcasts on standard frequencies of 2.5, 5, 10, 15, and 20 MHz.

The messages are changed each day at 0400 GMT, with provisions to schedule immediate alerts for outstanding occurring events. Geoalerts tell of geophysical events affecting radio propagation, stratospheric warming, etc.

SECTION 8

Interference

71

Sunspots and Blackouts

Mysterious dark areas on the sun's surface appear and disappear for periods of a few days to several months. Known as "sunspots," they seem to be whirling gases at a temperature of about half that of the remaining solar surface. Cooler temperature imparts the darker appearance. Although little is known about the origin of sunspots, they have a profound effect on shortwave radio.

When the number of sunspots is averaged, it is found to rise and fall over an 11-year cycle. A graph of this activity for 130 years is shown in Fig. 8-1. During a time of rising sunspots, the earth's ionosphere, which carries most long-range radio communications, experiences an increase in electrical activity. The total effect can be summed by a rule-of-thumb: the greater the number of sunspots, the shorter will be the wavelength of signals that skip great distances. In other words, a peak of the cycle is accompanied by generally improved radio conditions.

Scientists describe this effect as a rise in the MUF, or *M*aximum *U*sable *F*requency. This is the highest frequency that can return to earth after skipping through the ionosphere. For example, if you listen during a low period in the cycle, frequencies above 15 MHz may not be heard from distant countries. About six years later, when the number of sunspots reaches a peak, then frequencies as high as 35 MHz may be easily received between continents.

If you wish to anticipate radio conditions on the higher frequencies (usually above 15 MHz), then learn your position in

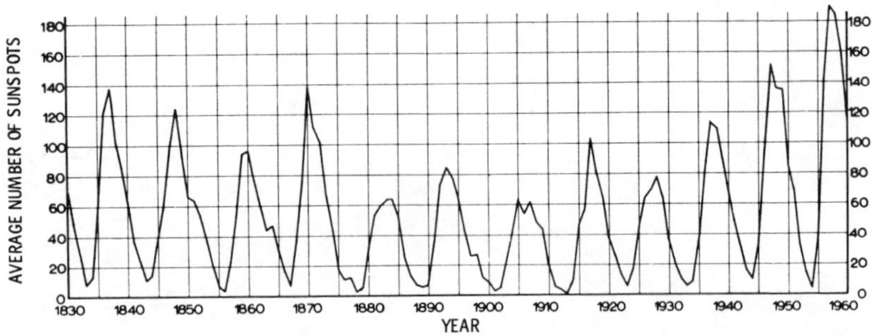

Courtesy National Bureau of Standards

Fig. 8-1. Variations in sunspot activity, 1830 to 1960.

the cycle. Low points are predicted for 1975 and 1986. Each low point is followed by increasing improvement in radio conditions as the number of sunspots rises toward the next peak about 5½ years later.

An increasing number of sunspots also brings an occasional radio "blackout." A brilliant flash erupts in the vicinity of a sunspot and widens over a large area of the sun's surface. Not long after, a distinct effect is felt on earth and a "solar flare" is observed. A huge quantity of ultraviolet and X-radiation bombards the ionosphere and may temporarily disable communications on the side of the world facing the sun. One of the worst blackouts on record was produced by solar flares over a 3-day period in 1957.

72

Noisy Brushes

Many appliances in the home are potential sources of interference to shortwave listening. There are electric mixers, power tools, vacuum cleaners, and oil-burner motors. In many cases, each produces interference that is easy to identify; the disturbance disappears when the motor is turned off. Some cases, like the oil burner, are not practical to treat, but serious interference can be cured by calling a burner serviceman. He may have special bypass devices made especially to fit an oil burner to cure its annoying variety of electrical hash. This interference is caused by a high-voltage discharge, as in an automobile spark plug.

Other electrical items in the home generate a different type of interference. A common one is sparking at the carbon brushes in brush-type electric motors, and a home handyman can often treat the problem. (Not all electric motors have brushes, however.) Leading causes of brush sparking are ordinary wear and dirt. Be sure the appliance plug is removed from the wall socket, then open the case to locate the brushes. After the brushes are removed, you should be able to see a ring of copper bars that form the commutator (see Fig. 8-2). Begin

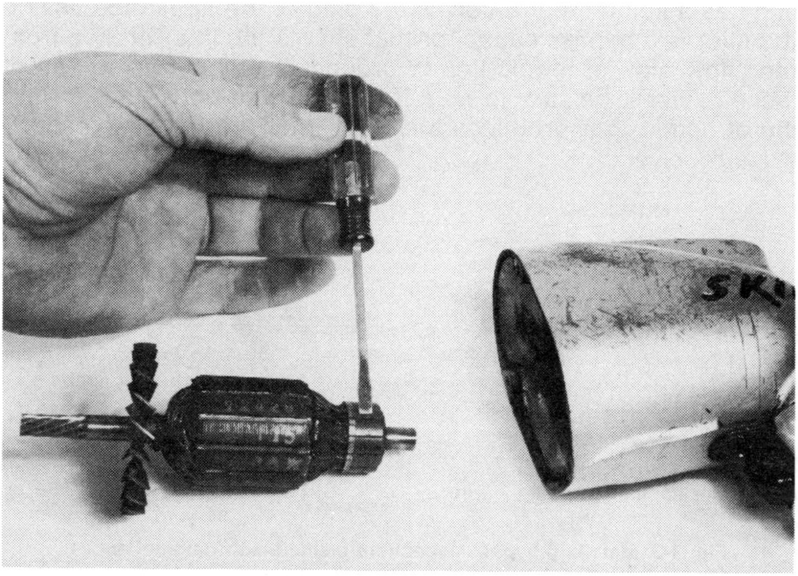

Fig. 8-2. Location of commutator on brush-type electric motor.

by blowing all dust and other foreign particles out of the appliance case. Since the commutator is probably blackened by the carbon brushes, use very fine sandpaper to shine the copper. You must not allow any metal dust or shreds to bridge across the bars of the commutator since these spaces must be electrical insulators. For this reason, never use emery paper for the cleanup.

If brushes have been in use for a long time, or have weak pressure springs, replace these items at an early date. Some alcohol on a cloth is helpful to remove grease and other matter that may interfere with motor operation. Cleaning up a motor will not only reduce radio interference but will provide you a bonus in improved operation of the appliance.

73

Appliance Hash

Simple filters are widely available that will minimize the "hash," buzzes, or other electrical interference caused by household appliances. The low-cost filter is usually enclosed within a small plastic housing that plugs into the ac wall outlet and has a socket that receives the plug of the appliance. Inside the filter is a bypass capacitor that shunts the interference from the "hot" side of the ac line to ground. If you purchase one of these devices, be sure to plug it into the wall both ways to find the direction that produces the least interference in the shortwave receiver.

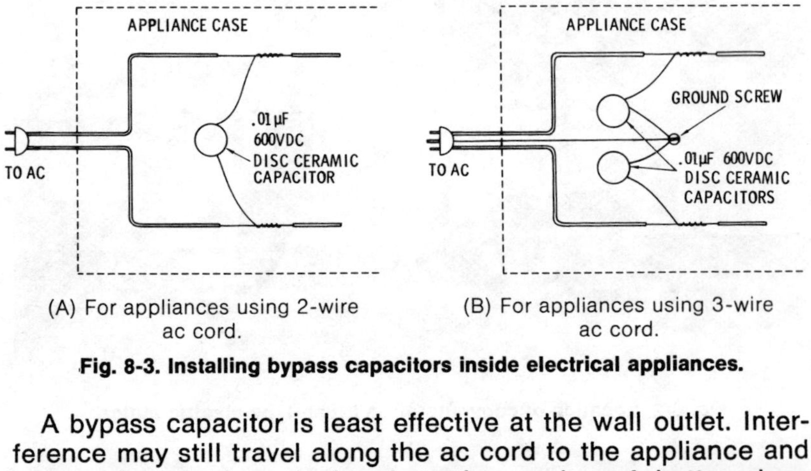

(A) For appliances using 2-wire ac cord.

(B) For appliances using 3-wire ac cord.

Fig. 8-3. Installing bypass capacitors inside electrical appliances.

A bypass capacitor is least effective at the wall outlet. Interference may still travel along the ac cord to the appliance and radiate directly through the air to the receiver. A better place for the capacitor is inside the appliance. Obtain a ceramic capacitor rated at .01 μF, 600 volts dc, and install it across the ac leads inside the appliance (see Fig. 8-3A). You will have to insulate any bare leads that might short circuit against the appliance case with electrical tape or spaghetti tubing. Also check to be sure that vibration developed when the appliance is operating will not damage the capacitor.

Many appliances have a three-wire ac cord and this calls for a second bypass capacitor of identical value. Connect the capacitors across each side of the ac line as shown in Fig. 8-3B. You can identify the ground lead since it is usually connected internally to the appliance case. All connections should be soldered and carefully taped for insulation and physical protection.

Bypass capacitors alone may not cure interference developed by office equipment such as adding machines and cash registers. These machines are more successfully treated with an inductance-type line filter. Such filters contain rf choke coils, as well as bypass capacitors, and do a better job of eliminating electrical interference. A good ground is important for proper filter action in these devices.

74

ITV

The letters "ITV" mean "Interference *from* Television," and represent a serious problem to shortwave listening in some locations. The trouble stems from the horizontal scanning circuits contained within every television receiver. These circuits generate a signal at a frequency of 15,750 Hz, and it is this signal that produces the "raster" of horizontal scanning lines across the face of the tv picture tube. Strong currents at the horizontal scanning frequency are applied to deflection coils surrounding the neck of the picture tube. These coils act as an antenna and radiate a considerable amount of electromagnetic energy into the area around the tv receiver. Although the frequency is well below the lowest bands of a shortwave receiver, the nature of the scanning waveform (sawtooth) is such that it is extremely rich in harmonic energy. The sawtooth waveform creates a multitude of harmonic signals (multiples of 15.75 kHz) that extend well up into the shortwave bands.

Interference from television has a distinctive sound. It is a sort of raspy buzz that also has a kind of musical sound. When you rotate the dial on the shortwave receiver, the interference will repeat itself at close intervals along the band. This indicates that you're hearing a succession of increasingly higher harmonics.

There are several "cures" for ITV, and some trial and error may be needed to find the ones that work. The simplest method, if you're fortunate, is to increase the distance between the tv and shortwave receivers. Sometimes just a small difference in spacing will make a big difference in the amount of interference. Anything that improves the signal strength to the shortwave receiver (such as a better antenna) will improve the resistance of the receiver to ITV. An external antenna leading away from the tv receiver might do it, especially if the lead-in is coaxial. Coaxial cable is less sensitive to interference pickup.

Another possibility is to try plugging the shortwave receiver into a different ac outlet. The receiver may be sharing a branch circuit with the tv and be suffering direct interference through the power line. If your receiver is a solid-state type, try operating it from batteries. This can vastly cut down interference arriving through the ac line. Be sure that both tv and shortwave receivers are equipped with line filters as described in Item 73.

If none of the preceding works, here is what some experimenters have tried, with varying degrees of success, when the tv receiver does not have a metal cabinet. It requires removal of the tv chassis, so don't attempt the job unless you have had some experience with it. After the chassis is removed, metal screening is stapled or fastened to the inside surfaces of the tv cabinet. After the screening is mounted, the pieces must be bonded together to ensure a continuous electrical connection throughout. A ground wire is connected to the screening, then run to a good electrical ground (such as described in Section 1). One precaution is to make sure that the screening does not cause short circuiting of any components on the tv chassis after it is replaced inside the cabinet.

As a last resort (which should be obvious), ITV can be completely eliminated by simply turning the tv off when listening to shortwave.

75

Fluorescent Lights

Fluorescent lighting in the home can be a prolific generator of noise in a shortwave receiver. Unlike filament-type lamps, which merely heat up and glow, fluorescent lamps create a rapid discharge of electrical current through ionized gas at the rate of 60 times per second. The steepness of the discharge sets up radio waves that are rich in harmonic energy. The interference is heard in the speaker as a rough buzz. The listener is usually aware of the source since it commences or stops when the light switch is operated. The disturbance reaches the receiver either through direct radiation or through the power line.

The first precaution is to locate the shortwave receiver as far as possible from any fluorescent lamp known to be producing interference. If your receiver is a solid-state portable, it can be operated on its internal batteries. This isolates it from interference arriving from the ac line.

If none of these steps improves the condition, try to reduce the interference at the source. Often, the most effective method is to install bypass capacitors inside the lamp fixture. After the power is removed, the fixture is disassembled, and two 0.1 μF capacitors rated at 600 volts dc are installed. One capacitor is connected between each side of the ac line and the fixture. (This assumes that conventional wiring has been used and that the metal fixture is electrically grounded.) Tape any bare wires to prevent short circuits and check to see that no wires are damaged when the fixture is reinstalled.

If the lamp is a portable type, try connecting a .01 μF, 600 V dc bypass capacitor across the ac line inside the lamp enclosure. After the capacitor is connected, try reversing the lamp plug in the ac wall outlet to determine which position produces the least noise in the receiver. Further quieting might be possible by installing line filters, as mentioned elsewhere in this section, on both the shortwave receiver and the fluorescent lamp.

76

Noisy Incandescents

This source of interference is not too common and thus rarely comes under suspicion. It is the ordinary incandescent light bulb, the filament type that is used in most homes. When an incandescent becomes noisy, it is usually a sign that it is about to burn out. The lamp filament develops a partial crack or break, across which electrical arcing occurs. The arcing produces radio-wave radiation that is heard as a crackling noise in the shortwave receiver.

The noisy light bulb can usually be located by switching the various bulbs in the house on and off. The interference will cease when the guilty lamp is turned off. The defective bulb will usually be located fairly close to the receiver, and sometimes it can be transferred to a more distant location in the home and continue to provide light for some time while not causing radio interference. In most cases, however, the best cure is to simply replace the noisy lamp.

In cases where the interference is intermittent, you may have to tap each lamp gently with the end of a pencil (Fig. 8-4). This will usually jar the lamp into revealing itself as a noise generator.

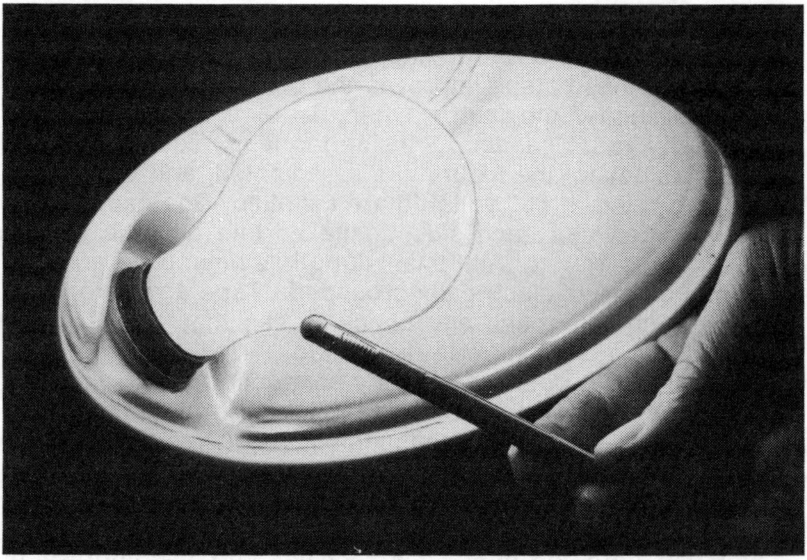

Fig. 8-4. Tapping with a pencil will often reveal noisy incandescent lamp.

77

Nearby Transmitters

The large number of transmitters now found in private homes (mainly due to the popularity of Citizens-band radio) increases the possibilities of interference. A transmitter in the next house or apartment might radiate an extremely strong signal into your shortwave receiver. Although these signals usually enter through the "front end," or rf section of the shortwave receiver, another common route is through the audio amplifier. You can usually determine this by noting whether the interference responds to changes in the receiver tuning dial. If it does not, it is being picked up in the audio section. The voices (or garbled sounds) are usually heard only briefly, while the transmitter is being operated.

Some possible routes into the shortwave receiver are the ac line cord, speaker leads, and chassis wiring. Many cases can be cured with bypass capacitors strategically placed in the receiver. Try a .01 μF, 600 V dc disc-ceramic capacitor across each leg of the ac line to the chassis. Install them at the point where the line cord enters the chassis.

Also, try a .01 μF disc-ceramic capacitor across the speaker leads. This is especially important if you are using an external speaker and the wires are long. The bypasses should be inside the chassis.

If these measures do not reduce the interference, another possibility is to bypass the input of the audio amplifier. A common entry point for radio-frequency interference is at the first audio amplifier. The tube or transistor at this point acts not only as an amplifier, but also as a diode detector which recovers the audio signal from the interfering radio wave. A small bypass capacitor can be connected here that will not affect audio frequencies, but will shunt the radio-frequency energy to ground. The usual point in the circuit is at the grid of the first audio tube; or, in the case of solid-state receivers, at the base of the first audio transistor. In most circuits, this point is easy to find because it is connected to the center terminal of the volume control. Find the control and connect a suitable bypass capacitor between the center control lug and ground.

The value of the capacitor will vary according to the type of circuit. In a tube receiver, a 50-pF mica capacitor should be approximately correct. In transistor circuits, 270 pF is the approximate value. You can determine the correct value by listening to the receiver while you temporarily connect different values across the volume control. The capacitance value must be large enough to bypass offending radio frequencies, but not large enough to reduce the treble or high-frequency response of the audio amplifier.

78

Utility Lines

The power lines that carry electricity through cities and into homes occasionally cause a kind of interference that disturbs shortwave reception. The direct cause is electrical leakage or sparking across insulators which may have become coated with airborne contaminants or are cracked. The electrical discharges create radio waves that comprise the interference. Sometimes a power transformer on a utility pole becomes defective and produces a similar effect. These problems may go unnoticed by the power company because they do not often impair the distribution of electrical power. Shortwave receivers, however, can suffer considerable noise interference. The noise is often intermittent and becomes worse when the weather is rainy or humid (because moisture helps the leakage path).

Since the same noise often affects a regular a-m radio, the radio becomes a handy test instrument for tracking down the noise source. Take a small a-m transistor radio outside, and, while carrying it about your neighborhood, try to determine if there is any pattern of signal strength to the noise. If the interference becomes stronger as you walk in a given direction, attempt to "home" in on it. You might identify a specific utility pole or area that contains the source of the noise. A call to the power company should clear up the matter in a reasonable time. Radio interference is an indication to the power company that something more serious—an interruption in electrical service—might soon occur because of defective insulators or transformers.

79

Noise Limiter

Impulse noise is a type of interference that is generated by automobile ignition systems, certain types of machinery, and neon signs, to name a few. Viewed on an oscilloscope, this noise is seen as a sharp spike which rises well above the desired audio program material. If your shortwave receiver has no built-in noise limiter to reduce such interference, you can assemble a simple one that works with earphones. With just a few components, it can remove impulse noises that often make listening impossible.

As shown in Fig. 8-5, the circuit consists of two 1N34A diodes, two 1.5-volt batteries (any size), a dpst switch for turning off the circuit when not in use, and a 5000-ohm level control. The circuit can be built inside the receiver in series with the wires to the earphone jack, or in a small external enclosure. Be certain

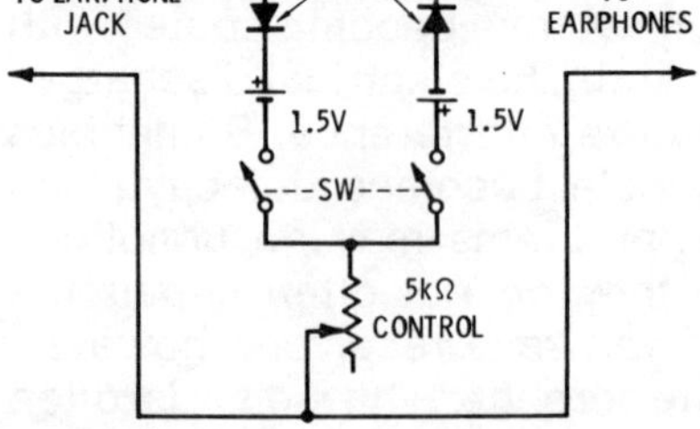

Fig. 8-5. Noise limiter.

to observe correct polarities of the diodes and batteries to assure proper operation.

The theory of the noise limiter is this: When the audio signal is less than 1.5 volts it cannot go through either diode to ground and proceeds in normal fashion to the earphones. Any signal that exceeds 1.5 volts, however, will pass through the diode circuits to ground. The excessive signal will be due to noise spikes and they will be shunted to ground. Two diodes, each connected in a reverse-bias condition but in opposite polarities, handle both positive and negative portions of the interfering signal.

In this circuit, noise is not completely eradicated but is reduced in level to the same strength as that of the desired audio. The limiting action, however, produces a significant quieting effect on noisy reception. The 5000-ohm control allows a measure of adjustment to match the circuit to different receivers.

SECTION 9

Troubleshooting

80

Slipping Dial Cord

When you turn a tuning knob, but the pointer fails to move across the dial, you have a case of slipping dial cord. This may seem like a harmless, easily treated trouble, but most service technicians have spent more hours on it than they care to remember. If your receiver develops this problem (and it probably will sooner or later), here are some tips on prevention and cure.

A simple dial-cord drive is shown in Fig. 9-1. Although many systems are somewhat more complex, this will serve to illustrate the basic features. There is a small shaft (at right) that is turned by a front-panel knob. The shaft rotates a large drum (left) that is mounted on the variable tuning capacitor. Somewhere along the dial cord is a pointer that indicates frequency on the front panel. Note that a tension spring exerts a pull on the cord at point "E."

The first sign of trouble is usually slippage when the pointer is at one end of the dial; it will return only after some jiggling of the tuning knob. When this symptom first occurs, it is usually best to do nothing more than purchase a small jar of dial-cord "dressing." Dab some dressing along the cord, around the tuning drum, and especially on the shaft (point "C"), where greatest slippage often occurs. This should cure most early cases of dial-cord slippage.

The dial cord will inevitably slip again at some future date. If the nonslip dressing is no longer effective, you will have to try another measure. Examine the tension spring (point "E") to see if it has lost tension. A completely closed spring means that

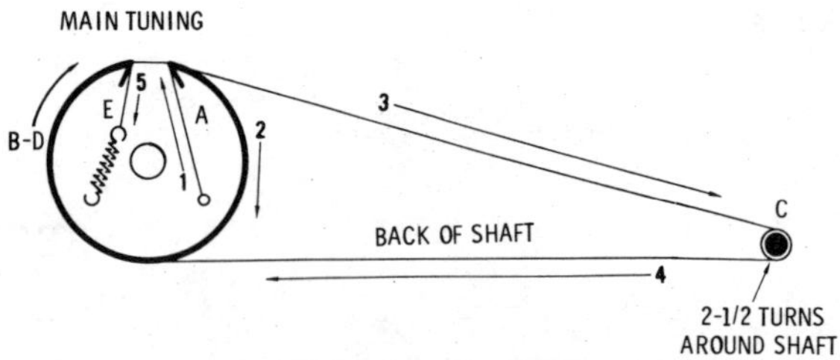

Fig. 9-1. Typical dial-cord stringing diagram.

the dial cord has stretched. This can be remedied by *slightly* shortening the cord where it is attached to the spring at point "E." Be careful to *not* remove the cord from the drum, and do *not* overstretch the spring, as this could cause the spring to lose tension permanently or pull the cord so strongly that the drive system binds. The spring should simply tug gently on the end of the cord.

After you shorten the cord, always run the pointer all the way across the dial and back several times. You may discover that the spring has lost tension again, and the cord needs a bit more shortening.

Only as a last resort, or in cases of dial-cord breakage, should you attempt to replace the dial cord. If you must do this job, you must have two essential items: the manufacturer's stringing diagram (such as the one shown in Fig. 9-1), and *genuine* dial cord, since any other kind of cord will quickly stretch. Two other items are very helpful. One is another person, since the job is often a three-handed one. Another is a piece of stiff wire about six inches long, with a tiny hook bent into one end. This tool can thread the cord into corners and crevices that are too small for your fingers to reach into.

81

Defective Speaker

Loudspeakers are remarkably durable, but they are not completely trouble-free. A burned-out voice coil is extremely rare in shortwave receivers since the audio amplifier does not generate sufficient power (as in hi-fi equipment) to melt the voice-coil wires. An open voice coil can be easily detected by placing the probes of an ohmmeter across the speaker leads. The speaker should produce a "click" when the battery voltage applied by the meter sends a surge of current through the coil and causes

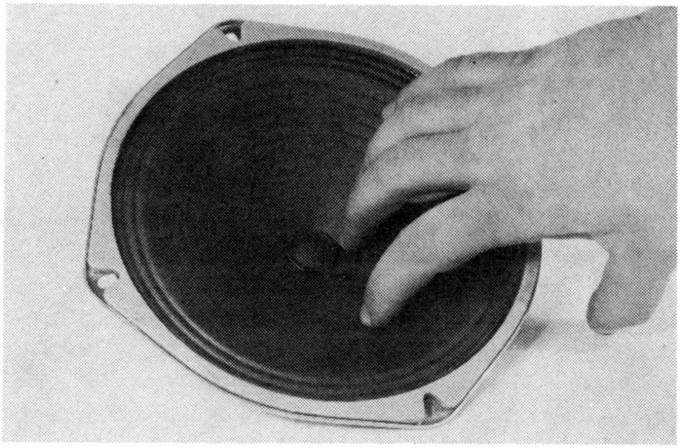

Fig. 9-2. Testing speaker for rubbing voice coil.

the cone to move. The meter should indicate a few ohms resistance if the voice coil is good.

A more common trouble is a misaligned speaker cone. After some time in operation, the cone suspension may go awry and cause rubbing between the voice coil and the permanent-magnet pole piece over which it moves. The usual symptom here is audio distortion that often sounds like a defect in the audio amplifier stages. The distortion may be heard continuously, but it usually sounds worse on certain musical notes or voice tones. A simple test on the speaker can expose this trouble very quickly. Place your fingers near the center of the cone, as shown in Fig. 9-2, and gently press inward. You should feel no rubbing or roughness, but only a slight spring-like tension. Sometimes you can hear the sound of the voice coil scratching against the magnet. If any of these symptoms appear, replace the speaker with a new one.

82

Fuzzy Audio

A torn speaker cone is a prime suspect in cases of fuzzy-sounding audio. This trouble is often caused by mishandling when the radio chassis is removed from the cabinet for repair, or when a sharp instrument has been pushed through the speaker grille cloth. The ripped edges of the paper cone produce the distorted sound. It may appear on all frequencies, or just on certain tonal combinations. This is one repair job that

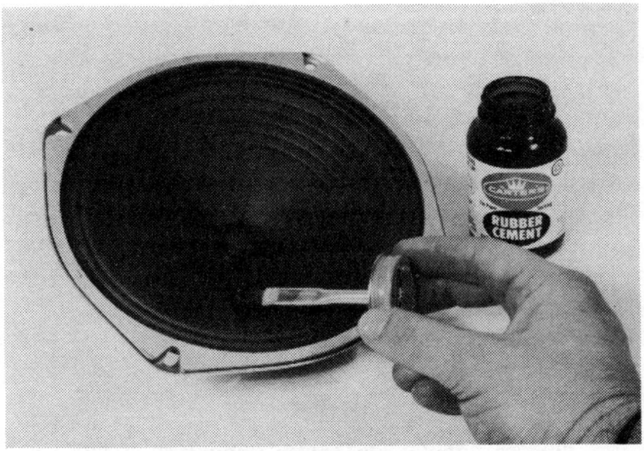

Fig. 9-3. Using rubber cement to repair torn speaker cone.

is often easy to perform. Just be certain to obtain a nonshrinking type of cement. Regular glue is apt to cause the paper cone to wrinkle. You can use "service" cement, or some other adhesive especially intended for speaker repair.

Another possibility, if no electronic supply houses are nearby, is to use rubber cement, which is sold in stationery stores (Fig. 9-3). If you believe you already have a suitable type glue, but wish to verify that it will work, try this simple test. Tear a piece of writing paper, then join the edges with the glue. After the glue has dried, there should be no sign of wrinkling or other stress in the paper.

83

Keep Cool

The holes you see in most shortwave receiver cabinets are designed to keep the circuit components ventilated. There may be perforations in the cabinet (Fig. 9-4A), or slots or holes cut into the rear cover (Fig. 9-4B). By allowing outside air to enter, they reduce an important cause of shortened component life—excessive heating. The air circulation is based on the convection principle. Heat developed in the chassis warms the air inside the cabinet. The heated air rises and exits from holes at the sides or top of the cabinet. Since this creates a partial vacuum within the cabinet, cooler air is drawn through the holes

(A) Holes in top of cabinet allow heated air to escape.

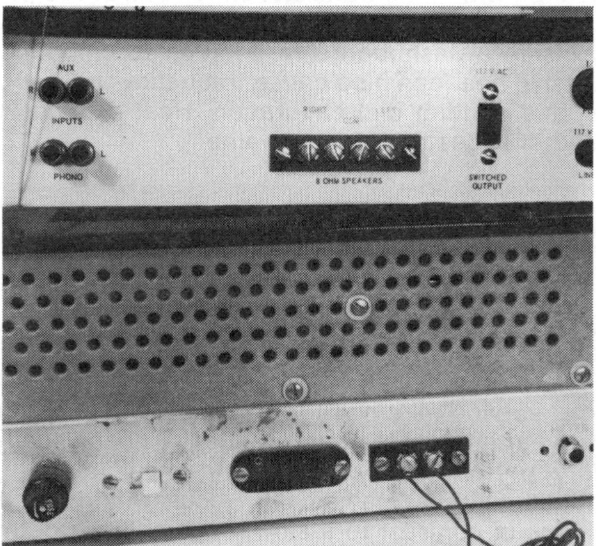

(B) Holes in rear cover allow cooler air to enter.

Fig. 9-4. Holes in receiver cabinet provide convection cooling of circuit components.

in or near the bottom of the cabinet, and the components are maintained at a relatively low temperature.

Since this simple form of air conditioning must be complete and continuous, any blockage in the flow of air raises the internal temperature and shortens component life. It is advisable to allow at least one inch of outside clearance for the openings. Do not allow books, papers, or other items to restrict the air flow (Fig. 9-4A). If you stack different pieces of electronic equipment vertically, watch out for troublesome combinations such as a receiver placed above another piece of electronic equipment that gives off excessive heat (Fig. 9-4B). Never place a receiver on a radiator or similar heat source. If you mount a receiver under the dash of an automobile, keep it out of the air blast of the car heater.

Precautions about air flow apply to solid-state equipment as well as to tube equipment. Although solid-state equipment does not generate nearly as much heat as tube equipment, certain semiconductors, especially those in a power supply, as well as other parts (filter resistors, for example) generate some heat which must be dissipated. If a solid-state circuit is not properly protected against heating, certain transistors can be triggered into a condition called *thermal runaway.* Heat build-up touches off self-heating within the transistor which can cause permanent damage. (This was a notorious problem in early mobile radios mounted inside a dashboard.) In addition to component damage, excessive heat can also cause frequency drift, poor tuning accuracy, and shifting dial calibration. Heat expands the wire in coils and changes their tuning value.

84

Check Tubes

How often should you check the tubes in a shortwave receiver? There is no prescribed time limit, but equipment in commercial service may call for an annual checkup. For home use, a two-year interval is more practical. Tubes usually deliver satisfactory performance well beyond their rated lives, and many sudden failures statistically occur within the first 100 hours. This is in contrast to a slow, accumulative deterioration that reduces, rather than interrupts, circuit operation. After some five or more years of typical home service, a vacuum tube filament may simply rupture under the repeated thermal assault of heating and cooling.

Certain tubes tend to suffer a noticeable loss of electron emission. These are usually high-current types, such as the rectifier or audio-output stage. This may lead to diminished audio output or a shift in circuit voltages. Tubes in the rf and i-f stages may develop high-resistance shorts between heater and cathode or between other internal elements. This can cause an increase in hum, or detuning.

When testing tubes, the simple tube checkers are mainly good for determining open filaments, large drop in cathode emission, or dead shorts. Less obvious problems such as low amplification, leakage, gas, etc., require a more sophisticated instrument such as the mutual-conductance tube checker found in most service shops. Another way to test tubes is through the technique of substituting a known good tube for a suspicious one and noting any improvement.

85

Howling Speaker

An unexplained whistle or howl in a loudspeaker is often caused by *acoustic feedback.* This is a path through the air or chassis that sends mechanical vibrations to an early stage in the circuit. The vibrations are converted to electrical signals that are amplified, then returned once again by the speaker. A build-up of these disturbances through the enclosed pathway produces the characteristic howl or oscillation. It can be especially annoying when a receiver is tuned to shortwave bands, because the path is more easily completed at higher frequencies. There are several possible causes and cures for this trouble.

A leading troublespot is the variable tuning capacitor (Fig. 9-5A). The feedback causes the thin plates of the tuning capacitor to vibrate and produce an electrical signal. Closely examine the capacitor mounting to be sure that no hardware is loose and that rubber dampers (if used) are still performing their job. Another troublespot is the large dial-cord drum that often rides on the tuning-capacitor shaft. The drum can act as a collector of sound waves from the speaker and transmit vibrations to the tuning-capacitor plates. If the drum appears to be the culprit (you should be able to vary the symptom by pressing on the drum), check to see if the drum's mounting screws are loose. If tightening the screws does not effect a solution, glue pieces of felt on the drum to dampen the acoustic response.

(A) Tuning-capacitor plates and dial-cord drum.

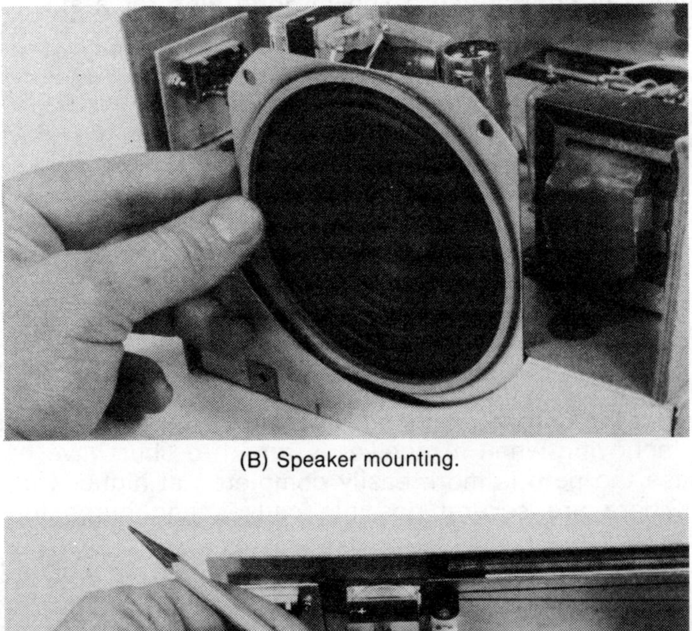

(B) Speaker mounting.

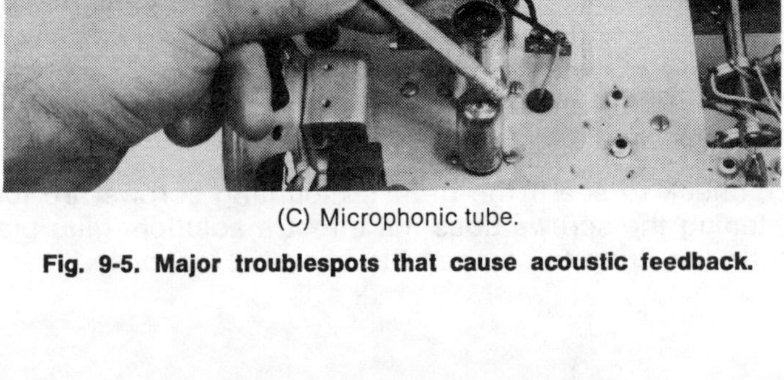

(C) Microphonic tube.

Fig. 9-5. Major troublespots that cause acoustic feedback.

Sometimes a speaker can radiate vibrations directly to the plates of the tuning capacitor. These cases can often be cured by placing an acoustic shield between the speaker and the capacitor plates—a sheet of material coated with absorbent material like felt, for example. If such treatment does not stop the vibrations, try remounting the speaker to soften the direct connection between the speaker frame and the receiver cabinet (Fig. 9-5B). Rubber or felt material placed around the edge of the speaker may sufficiently interrupt the feedback path.

Some receivers avoid the feedback problem by not supplying a speaker inside the receiver cabinet. The speaker is sold separately. In this case, do not place the speaker on top of the receiver where a direct physical connection can touch off the howl.

A common cause of speaker howling is a "microphonic" tube. The elements within a vacuum tube may loosen with age and commence to shake in step with speaker vibrations picked up through the chassis. This can often be determined by tapping each tube in the circuit with a pencil eraser to locate the sensitive one that causes the howl (Fig. 9-5C). Replacement of the tube with a new one usually cures the trouble.

86

Noisy Volume Control

One trouble you can anticipate in an aging shortwave receiver is physical wear in the volume control. The control has a metal wiper arm that presses against a circular carbon element. With the passage of time the wiper loses tension, and this gives rise to several symptoms. One is a scratching sound that is heard as the volume-control knob is rotated. This is caused by reduced pressure of the wiper arm against the carbon element. In other cases, the arm wears a "dead" spot in the carbon element because the volume control is usually left on a favorite listening level. This can cause complete loss of audio, or intermittent sound which cuts the audio on and off as someone walks across the room. Another common symptom is hum in the audio caused by excessive contact resistance between the wiper arm and the carbon element. Some vigorous turning or tapping of the control knob while listening will pinpoint the problem.

If the trouble is a mild one, a good cure is to spray the control with a contact cleaner (Fig. 9-6). Be sure the ac power is off and turn the control knob quickly back and forth as the spray enters.

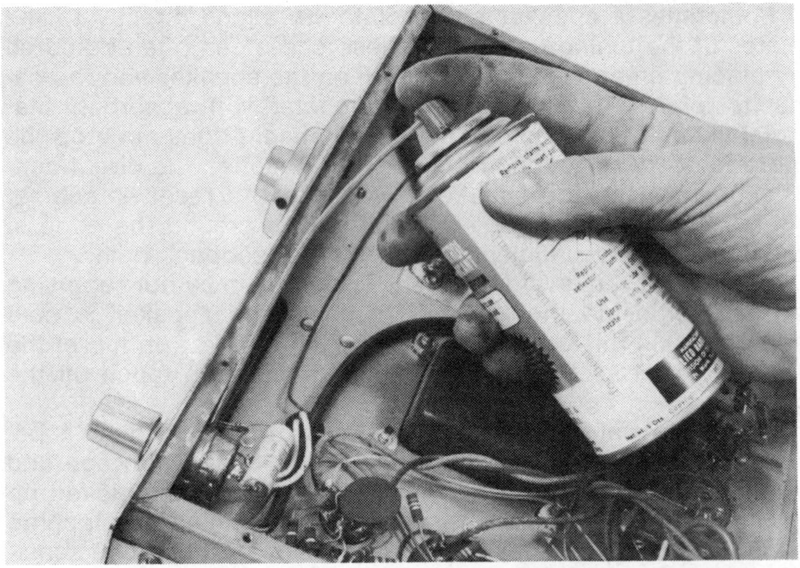

Fig. 9-6. Cleaning a noisy volume control with contact cleaner.

You will not have to remove or disassemble the control since there are openings around the terminal lugs through which the spray can be directed. An extension tube supplied with many aerosol cans allows you to direct the spray into tight spaces.

If cleaning the control does not cure the problem, you will have to replace it. Be sure to obtain an exact replacement control if possible. The job is easily done if you carefully tag the five or so wires soldered to the old control, and rewire them correctly to the new part.

87

Dirty Bandswitch

The bandswitch in a shortwave receiver is a critical item that will need attention after several years of use. Airborne dirt and grease reach the contact surfaces of the switch and produce symptoms that are easy to identify. One symptom is a station shifting in and out of tune when the bandswitch knob is tapped with a finger. Another symptom is a station that is tuned in perfectly may not remain in tune when the bandswitch is clicked on and off the band position. Noisy reception and loss of sensitivity are other indicators of a dirty bandswitch.

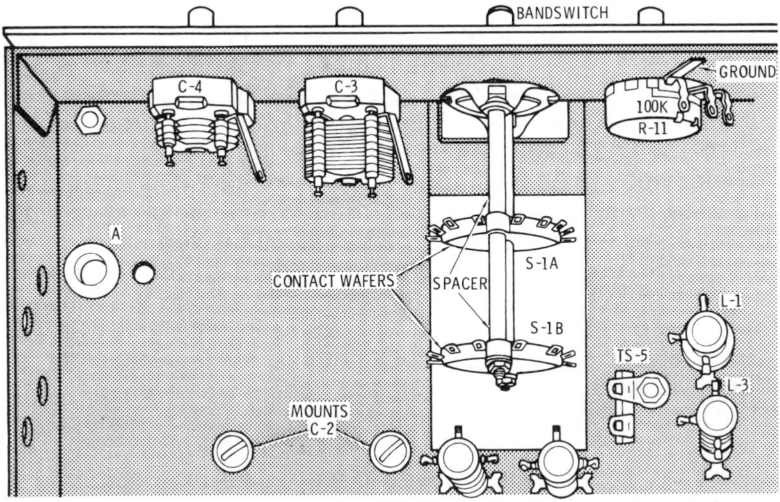

Fig. 9-7. Sketch showing the location of the bandswitch contact wafers.

The spring contacts on the switch may lose tension with the passage of time and no longer grip the switch poles tightly. Since replacement of a complete bandswitch is a delicate job and too critical for a casual hobbyist, try the next best treatment. Obtain an aerosol can of contact cleaner and spray the switch wafers on each side. These wafers, shown in Fig. 9-7 as S-1A and S-1B, lie in the back of the bandswitch knob. (There may be more wafers than are shown here.) The extension nozzle sold with contact cleaners is very helpful for getting into narrow spaces, and should be used.

As you spray the cleaner onto the switch wafers, keep turning the bandswitch knob back and forth to be sure the cleaner reaches all contact areas. Also, be careful not to move any wires connected to the wafers since they are part of the tuning circuits on the higher bands.

88

Squealing Batteries

If you hear whistling sounds behind each station you tune, you may be a victim of the "squealing" battery. These whistles are often a signal that the battery is near the end of its useful life. As batteries age, the audio volume drops because of reduced voltage, but this is also accompanied by increased resistance within the cell. This resistance appears between the battery terminals and causes instability in the amplifier stages. Signals that flow freely between these points when the battery is fresh, will be partially obstructed by the increased resistance and may cause the i-f amplifier to oscillate. This produces the whistles behind each station.

The obvious remedy is to replace the battery. One temporary cure used by some thrifty SWLs, though, is to connect an electrolytic capacitor across the plus and minus terminals of the battery. This provides an unobstructed ac path for signals that cause the oscillation (as does a fresh battery), and the whistling stops. The battery, of course, will expire in time and will have to be replaced since it cannot continue to develop sufficient voltage when its internal resistance rises excessively.

The value of the electrolytic capacitor should be about 20 μF or more. The working voltage should be greater than that of the battery; for example, in a 9-volt circuit, use a capacitor of 15 volts dc or higher. Be sure the capacitor polarity agrees with that of the battery. Although you can connect the capacitor directly across the two leads that emerge from the battery, there is a better way. Find the battery lead that goes to the on-off switch, and connect the capacitor lead (of the correct polarity) to the switch terminal that is away from the battery. This prevents the battery from continuously applying voltage to the capacitor when the radio is switched off.

If your receiver has batteries with friction-type connections (where the battery terminals merely press against each other), a similar problem (squealing) may develop. Corrosion may form, or insufficient pressure may introduce a small amount of resistance between the battery contacts. If you suspect this, remove the batteries and briskly rub their terminals with a cloth until they shine.

SECTION 10

Miscellaneous

89

Predicting Long-Distance Skip

On a day-to-day basis, the shortwave listener will increase his familiarity with how the various bands perform. Different regions of the world are heard at different times of the day, and on different frequencies. The experienced SWL is aware of seasonal changes that bring increased interference on certain bands, or the 11-year sunspot cycle that produces tremendous changes in long-haul signals. Although the ionosphere, which accounts for skip transmission, is under constant surveillance, its operation is not completely understood. Nevertheless, great strides have been made in mapping these electrified layers and recording their behavior.

It is possible to predict with reasonable accuracy how the ionosphere may be expected to perform three months in advance. This is done by adding the diverse factors that contribute to the makeup of the ionosphere—like time of day, season, and year—then feeding the data into computers which come up with a projection. This information is invaluable to common carriers, military services, the press, and anyone else who conducts communications through the ionopshere. International shortwave broadcasters also depend on these predictions.

The predictions are also available to any shortwave listener. Much of the data is complex and technical, but the serious student of shortwave broadcasting might be interested in the medium beyond its hobby aspect. For example, the world map shown in Fig. 10-1 is an actual prediction chart for one month.

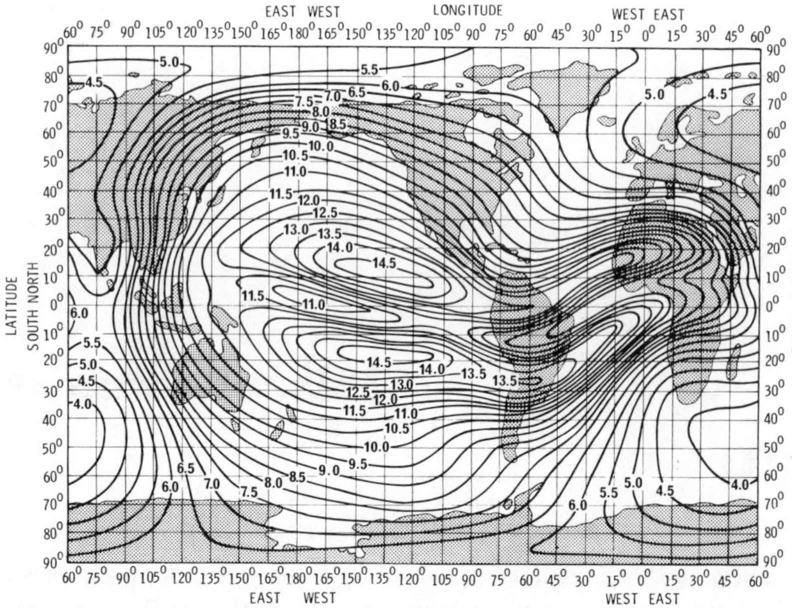

Fig. 10-1. Monthly ionospheric prediction chart.

It shows the highest frequency that can be used for successful communications over various paths throughout the earth. Such charts are issued on a monthly basis and are called "ITSA Ionospheric Predictions." If you're interested in obtaining them, first secure a copy of "Handbook for CRPL Ionospheric Predictions." This is a guide which explains (at a cost of less than a dollar) how the charts are used. Also known as "Handbook 90," the guide can be ordered by writing to: Superintendent of Documents, U.S. Government Printing Office, Washington, D.C. 20402.

90

Free Propagation Forecasts

A short-term prediction of radio conditions is broadcast in voice on each of the standard frequencies of the National Bureau of Standards (WWV). The frequencies are 2.5, 5, 10, 15, 20, and 25 MHz. The prediction is transmitted during part of every fifteenth minute of each hour. The announcement informs the listener of the ionospheric conditions during the next six hours.

Courtesy National Bureau of Standards

Fig. 10-2. Radio observatory in Lima, Peru, probes the ionosphere.

The information is collected from observatories (see Fig. 10-2) throughout the world, from radio soundings, solar activity, and similar information. The WWV broadcasts refer to routes in the North Atlantic area, such as Washington, D.C. to London, and New York to Berlin.

91

Fading

The strength of signals arriving from great distances through the ionosphere is rarely constant. Instability or changes in the electrified layers of the ionosphere can absorb signals or alter their reflecting angles. Signals may divide and travel different paths to the receiving antenna. These are common causes of *fading.* Little can be done about these effects other than changing frequency or awaiting improved band conditions. One technique used by commercial and military stations to counter fading is "diversity" reception. This requires two receiving antennas placed well apart, and two receivers. When the signal fades in one antenna, it may be satisfactory in the other. This approach, however, is usually too costly for the SWL hobbyist.

There is one variety of fading that can be treated with varying degrees of success. Known as "selective fading," it is heard as a severe distortion in audio quality that comes and goes. It may be worse on music than on voice. The cause is a critical condition in the ionosphere that bends each frequency of a given signal to a slightly different degree. A program carrying music, for example, carries three distinct frequency components—the lower sideband, the carrier, and the upper sideband. These encompass a band of frequencies about 10 kHz wide. As this composite signal enters the ionosphere, its different frequencies reflect back at slightly different angles. This causes different arrival times at the receiving antenna and cancellation of some tones in the transmitted music or voice.

If your receiver is capable of single-sideband reception (ssb), or simply has a control for varying selectivity, you can often improve reception. By slicing away part of the signal (only one sideband is actually required), you may reduce the distortion caused by selective fading.

92

TV DX

One branch of the SWL hobby specializes in finding distant television stations. Although tv signals are mainly in the vhf and uhf bands (from 54-216 MHz for vhf, and 470-960 MHz for uhf) and travel line-of-sight, there are temporary ionospheric conditions when tv stations from beyond the horizon can be received and logged. The specialist in this game of tv DXing will even photograph the screen to prove his catch.

The transportation medium for most tv DX is the "Sporadic-E" layer. This refers to a layer in the ionosphere whose appearance is unpredictable. When it materializes, it takes the form of small patches of very dense ionization. These electrified clouds can reflect signals much higher in frequency than those reflected by the conventional ionospheric layers. Sporadic-E pathways can often carry tv signals 500 miles or more.

The best time to look for tv DX is during spring and early summer, when these curious ionospheric layers tend to form. Try tuning between your local channels since they may block the distant arrivals. Also, tv stations are usually assigned alternate numbers on the dial to avoid mutual interference. This tends to place DX stations on unused channels in your area.

If you're really enthusiastic about seeing tv DX, try tuning when local stations are off the air (in the wee hours of the morning, or after evening sign-off). When the strong stations are shut down, you may be surprised by the faraway images you see when band conditions are right. Reception is greatly improved if your antenna system has a rotator. The DX station probably will not lie in the same direction as the locals, and antenna orientation can strengthen weak DX signals.

An excellent clue to good tv DX conditions is how you're receiving local stations. Every few years, the newspapers or radio warn about widespread interference that blanks out millions of tv screens. What is happening is that Sporadic-E transmission is bringing in distant signals that conflict with the locals. The classic symptom of this is the "venetian-blind" effect caused by two signals competing for the same channel. This produces coarse horizontal lines in the picture, and the audio takes on a characteristic interference that sounds like the grinding noise of a key being cut. (This is a beat note between the picture-synchronization pulses from the two stations.) The channels most affected by this skip condition are the lower ones from 2 through 6, although higher frequencies (even uhf) are sometimes affected also.

93

Canadian Time Station

The correct time is accurately announced every minute in GMT by the National Bureau of Standards stations WWV and WWVH (frequencies are given elsewhere.) These signals contain a considerable amount of information about frequency, pitch, propagation forecasts, geoalerts, and other references. But if you simply want the correct time, it is often more convenient to tune the Canadian time-signal station, CHU, operated by the Dominion Observatory, Ottawa, Ontario, Canada. Instead of time announcements in GMT, the Canadian station broadcasts it every minute in EST (Eastern Standard Time). Thus, you have less converting to do to find your local time.

The frequencies of CHU are 3330, 7335, and 14,670 kHz. These frequencies are spaced far enough apart on the shortwave bands so that you should be able to pick up at least one signal. You can identify the signals of CHU by a steady transmission of 1000-Hz tones, or beeps, spaced at 1-second intervals (as opposed to the clocklike ticks transmitted by WWV).

Ten seconds before each minute ends, the time is announced by voice; once in English, once in French. (The next minute, the announcers change the order and French is first.)

94

Entering Ham Radio

If you're tired of being on the receiving end all the time and want to talk back, then become a ham! It's a logical step from the ranks of the SWL into amateur radio. A good SWL receiver can be one-half of a ham rig (the most expensive part, too). Many communications receivers are calibrated for the ham bands or have bandspread dials that can be set up for ham operation. Major ham bands that can be tuned by the typical shortwave receiver are (in meters) 160, 80, 40, 20, 15, and 10. They display the same long-distance skip as do the international broadcasters.

Much of the experience acquired in SWLing can be transferred to amateur operation (which requires listening more than half the time). You'll copy stations through difficult interference and poor band conditions, suffer fades and other caprices of radio transmission. An ability to skillfully handle a receiver is a tremendous asset.

The easiest route to ham radio is the Novice license. Although its privileges are limited, the code requirement is modest (only 5 words per minute) and the written test is an uncomplicated mutiple-choice exam. Nevertheless, Novices operating with 75 watts of transmitter power have contacted more than 100 countries. As the Novice operates on the air under a limited license, he quickly builds code speed and the background to attain a higher-grade ticket with increased operating bands and privileges.

The Novice exam is obtained through the mail and administered by a local amateur operator (who also gives the code portion of the exam). If you don't already know a local ham, inquire at a local electronics supply store or ask a tv serviceman. Either source should know of a nearby ham who will be happy to help someone enter the interesting hobby of two-way shortwave.

95

Fail-Safe Insulator

Any outdoor antenna should be securely supported to prevent being a hazard to people below. Also, if the wire falls across utility lines, there is danger from electrical shock. One device to improve antenna security is the "egg" insulator, so-called because of its slight resemblance to an egg (see Fig. 10-3). The two pairs of holes for the supporting and antenna wires are arranged to form loops which run through each other, but are electrically insulated. If the insulator is broken, the antenna cannot come down because the loops lock together. This fail-safe feature should be considered if you have a long, heavy antenna in a crowded area.

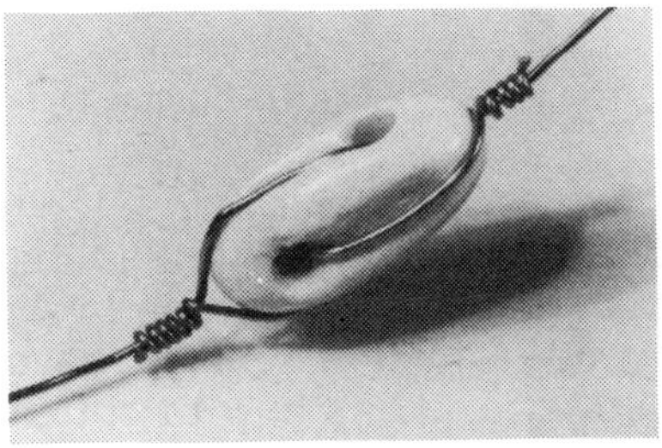

Fig. 10-3. An "egg" insulator.

96

Surplus Literature

Browsing through electronic surplus catalogs offered to experimenters and hobbyists can unearth interesting items. The range of equipment on the surplus market runs anywhere from headphones to radar jammers. The shortwave listener might discover useful accessories such as antenna hardware, speakers, and test meters. Military receivers that cover every corner of the radio spectrum are also plentiful on these pages.

The attraction of surplus equipment is low price often coupled with excellent design. To gain these benefits, though, the buyer must be selective. Much of this equipment requires special power sources, such as 28 volts dc. The circuits often need conversion to house-current operation with the aid of surplus

Fig. 10-4. Some typical surplus literature.

conversion literature and technical knowhow in electronics. Surplus equipment is often tube-operated and bulky. Yet, there are bargains to be had, and valuable items for the SWL. Almost any newsstand magazine devoted to the ham or electronic hobbyist explains how to obtain surplus literature such as that shown in Fig. 10-4.

97

Shortwave on TV

The shortage of frequencies in the two-way radio bands has prompted government officials to seek new ways to accommodate the growing number of stations. One solution is to place police, fire, and other "land mobile" radio in the uhf television band. Most uhf channel positions on a tv receiver are unoccupied, which permits such sharing of frequencies between different services. (Actually, the frequencies between 806 and 890 MHz, which were formerly designated uhf television channels 70 through 83, have been reallocated from television broadcasting to the land mobile services, but certain existing tv transmitters—known as "translators"—will be allowed to continue operating on a secondary basis for an indefinite period.)

Since two-way radio on the uhf band uses fm signals, as does a tv station for the audio portion of the signal, it is possible to pick up the mobile radio on a regular tv receiver. In almost all cases, however, land mobile broadcasting on these uhf frequencies is restricted to larger cities where communications activity runs highest. If you occasionally turn your uhf-tv channel selector over its complete range, you may locate and hear these mobile stations.

98

Broadcast-Band DX

Anyone who listens to the regular a-m broadcast band after dark is aware that stations many hundreds, even thousands of miles away are sometimes heard "loud and clear." This has created an interesting division of shortwave listening known as "broadcast-band DX." It requires little in the way of specialized equipment (although a long antenna may help) and the reward could be QSL cards from distant areas.

The stations are heard for more than one reason. Several are *clear channel,* operating at high power (usually 50,000 watts), and their signals cover wide areas of the country. Their frequencies are "protected" by the FCC so they suffer little or no interference. Although a clear-channel station may sound slanted at local audiences, its signals fill the gaps in rural areas to assure that every part of the country enjoys at least some broadcast service.

The second reason for hearing distant broadcast stations is the condition of the ionosphere. This electrical mirror, which carries shortwave signals great distances, also exerts a daily influence on the lower-frequency broadcast-band stations. During daylight hours, the bottom of the ionosphere grows very low under the continuous bombardment of the sun's ionizing rays. Any signals from a broadcast station that strike the ionosphere are reflected earthward at a very shallow angle. It is so shallow, in fact, that the reflected signal never returns to earth and is lost to space. In the evening, however, the ionosphere retreats to higher altitudes as electrified particles recombine into a neutral state. At night, then, signals from the same stations reflect downward at a much steeper angle and return to the earth's surface.

The daily effect is that low-power stations, intended only for local listening, will skip great distances. If you prepare a com-

(A) Original Marconi station on Cape Cod.

(B) Exhibition site today.

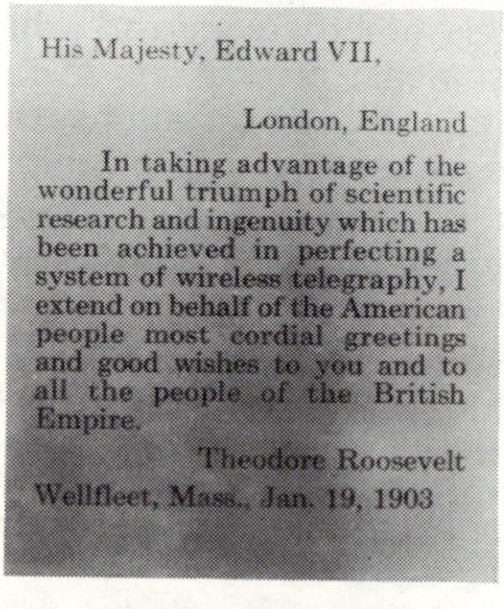

His Majesty, Edward VII,

London, England

In taking advantage of the wonderful triumph of scientific research and ingenuity which has been achieved in perfecting a system of wireless telegraphy, I extend on behalf of the American people most cordial greetings and good wishes to you and to all the people of the British Empire.

Theodore Roosevelt

Wellfleet, Mass., Jan. 19, 1903

(C) First U.S.-to-England message.

Fig. 10-5. Guglielmo Marconi's original wireless transmitter is historical site today.

plete reception report on what you hear, the distant station may reply with a valuable QSL card. Address your request to the Chief Engineer, in care of the station. During a station identification there is usually sufficient address given (call letters and town) for mailing purposes. If the state is not immediately apparent, listen a while until you hear it on a program such as the local news.

It's helpful to have a directory of stations, like the SAMS publication, *North American Radio-TV Station Guide,* by Vane A. Jones. This book lists all stations in North America, and provides other information such as power, type of programming, and hours of operation. One good tip for hearing new stations is to listen late at night at different hours. Stations don't sign off at the same time, and interference may be reduced on one you're trying to log.

99

Visit a Pioneer Shortwave Station

A fascinating exhibit maintained by the U.S. government is the ruins of the first Marconi station. Located in the town of Wellfleet, Mass., (near the tip of Cape Cod) it is worth a visit if you are in the area. Only remnants of the station exist today, but there is a scale-model replica on the site of the original installation.

As shown in the early photograph (Fig. 10-5A) four massive towers rose 210 feet high. They stood on a bluff overlooking the Atlantic Ocean, and commanded an unobstructed transmitting path to England. The station went on the air in 1903, barely two years after Marconi proved that shortwave signals could cross the ocean, and operated for 15 years. It communicated with ships at sea and handled overseas messages and news. As explained at the exhibit (Fig. 10-5B), the roar of electrical generators could be heard as far as three miles away.

The first message transmitted during the inauguration of the station was from President Theodore Roosevelt to the King of England. The historic text is shown in Fig. 10-5C.

SAMPLE STATION LOG

Date	Station	Location	Freq.	Time	Signal Report	Program Content	Remarks	Verification	
								Sent	Received

SAMPLE STATION LOG

Date	Station	Location	Freq.	Time	Signal Report	Program Content	Remarks	Verification	
								Sent	Received

SAMPLE STATION LOG

Date	Station	Location	Freq.	Time	Signal Report	Program Content	Remarks	Verification	
								Sent	Received